Design of Op-Amp Circuits, With Experiments

By

Howard M. Berlin

Originally Published as
The Design of Op-Amp Circuits, With Experiments
by E&L Instruments, Inc.

Howard W. Sams & Co., Inc.
4300 WEST 62ND ST. INDIANAPOLIS, INDIANA 46268 USA

4480538

International Standard Book Number: 0-672-21537-3
Library of Congress Catalog Card Number: 78-56606

Printed in the United States of America.

Preface

Perhaps the most widely used integrated circuit today is the operational amplifier. It is easy to use, requiring a minimum of algebra. This is probably why the present designs of many electronic circuits are simpler than they were a decade ago when some of us were still playing with vacuum tubes and transistors.

This book is about the design and operation of basic operational amplifier circuits, coupled with a series of over 35 experiments to illustrate the design and operation of linear amplifiers, differentiators and integrators, voltage and current converters, comparators, rectifiers, oscillators, active filters, and single power-supply circuits. It is not meant to be a sourcebook of all available operational-amplifier circuits, nor a textbook covering the performance characteristics of the various types that are available. However, this is a text/workbook that explains the designs of the fundamental circuits that are the building blocks of the more sophisticated systems using many operational amplifiers. For this reason, this book is useful to the beginning experimenter and hobbyist who wants to learn the basics by self-study, or it can easily serve as an addition to any college course on linear integrated circuits, especially those which have a laboratory section.

In nine chapters the fundamental circuits using bipolar and Norton-type operational amplifiers are discussed, with numerous numerical examples worked out. In addition, one chapter is an introduction to the increasingly popular instrumentation amplifier, which is useful for the amplification of low-level signals. The operational amplifier is perhaps the most versatile integrated circuit in use, which has led the authors of one book to proclaim: "Op-amps are the greatest thing since the contraceptive pill!"

As in my other books, I would like to thank David Larsen and Peter Rony, of the Virginia Polytechnic Institute and State University, the staff at Tychon, Inc. for their valuable advice and assistance, and E&L Instruments, Inc. who continue to support my efforts. In addition, I am grateful to the various manufacturers cited in this book for allowing me to reproduce technical data from their promotional literature and catalogs.

HOWARD M. BERLIN

Contents

CHAPTER 1

What Is an Op-Amp?

INTRODUCTION

What is an operational amplifier? In this chapter, we will define what an operational amplifier is, and discuss the many parameters that distinguish one type of device from another.

OBJECTIVES

At the completion of this chapter, you will be able to do the following:

- Define the following terms:

  channel separation
  closed-loop gain
  common-mode rejection ratio
  gain-bandwidth product
  input bias current offset
  input offset current
  input offset voltage
  input resistance
  input voltage range

  inverting input
  loop gain
  noninverting input
  open-loop gain
  operational amplifier
  output resistance
  output voltage swing
  slew rate

- Interpret a typical op-amp data sheet.
- Measure some of the common op-amp parameters.

THE IDEAL OP-AMP

Before we start looking at actual operational amplifier circuits, we will briefly consider the operational amplifier, hereafter referred to as the *op-amp*, by itself. The term op-amp was originally used to describe a series of high-performance dc amplifiers that were used as the basis for analog computers. Today's integrated circuit op-amp is a very high-gain dc amplifier that uses external feedback networks to control its response.

The op-amp without any external feedback is described as being used in an open-loop mode. It is in this mode that we can describe the characteristics of the ideal op-amp:

1. The open-loop gain is infinite.
2. The input resistance is infinite.
3. The output resistance is zero.
4. The bandwidth is infinite.
5. The output voltage is zero when the input voltage is zero (i.e., zero offset).

In practice, however, no op-amp can meet these five ideal open-loop characteristics. However, as we shall see in the next few chapters, the world doesn't come to an end because there is no such thing as the ideal op-amp.

THE OP-AMP SCHEMATIC SYMBOL

Depending on which magazine or book you read, the schematic symbol for the op-amp may have either of the two forms shown in Fig. 1-1. In either case, the op-amp has two inputs: one inverting,

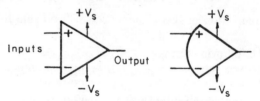

Fig. 1-1. Op-amp schematic symbols.

or − input, and one noninverting, or + input. Both have a single output. As shown in Fig. 1-1, the op-amp is powered normally by a dual-polarity power supply, typically in the range of ±5 to ±15 volts. However, it is the normal practice to omit these power supply connections, as they are implied.

THE OP-AMP DATA SHEET

Perhaps the best way to understand the many characteristics of an op-amp is to examine a manufacturer's data sheet, which can cover several pages in an integrated-circuits data manual. As shown in Fig. 1-2, the data sheet usually contains the following information:

1. A general description of the op-amp.
2. An internal equivalent circuit schematic.
3. Pin configuration of the device.
4. The absolute maximum ratings.
5. The electrical characteristics.
6. Typical performance curves.

In this section, we will cover most of the important parameters, using the type 741 op-amp as a representative example.

Maximum Ratings

The maximum ratings given in the data sheet (e.g., Fig. 1-2), are the maximum the op-amp can safely tolerate without the possibility of destruction.

1. *Supply Voltage* ($\pm V_s$)
   This is the maximum positive and negative voltage that can be used to power the op-amp.
2. *Internal Power Dissipation* (P_D)
   This is the maximum power that the op-amp is capable of dissipating, given a specified ambient temperature (i.e., 500 mW @ < 75°C).
3. *Differential Input Voltage* (V_{id})
   This is the maximum voltage that can be applied across the + and − inputs.
4. *Input Voltage* (V_{icm})
   This is the maximum input voltage that can be simultaneously applied between both inputs and ground, also referred to as the common-mode voltage. In general, this maximum voltage is equal to the supply voltage.
5. *Operating Temperature* (T_a)
   This is the ambient temperature range for which the op-amp will operate within the manufacturer's specifications. Note that the military grade version (741) has a wider temperature range than the commercial, or hobbyist, grade version (741C).
6. *Output Short-Circuit Duration*
   This is the amount of time that the op-amp's output can be short-circuited to ground or either supply voltage.

Electrical Characteristics

The op-amp's electrical characteristics are usually specified for a given supply voltage and ambient temperature. However, certain parameters may also have other conditions attached, such as a particular load or source resistance. Generally, each parameter will have a minimum, typical, and/or maximum value.

9

signetics

LINEAR INTEGRATED CIRCUITS

DESCRIPTION
The μA741 is a high performance operational amplifier with high open loop gain, internal compensation, high common mode range and exceptional temperature stability. The μA741 is short-circuit protected and allows for nulling of offset voltage.

FEATURES
- INTERNAL FREQUENCY COMPENSATION
- SHORT CIRCUIT PROTECTION
- OFFSET VOLTAGE NULL CAPABILITY
- EXCELLENT TEMPERATURE STABILITY
- HIGH INPUT VOLTAGE RANGE
- NO LATCH-UP

ABSOLUTE MAXIMUM RATINGS

	μA741C	μA741
Supply Voltage	±18V	±22V
Internal Power Dissipation (Note 1)	500mW	500mW
Differential Input Voltage	±30V	±30V
Input Voltage (Note 2)	±15V	±15V
Voltage between Offset Null and V⁻	±0.5V	±0.5V
Operating Temperature Range	0°C to +70°C	-55°C to +125°C
Storage Temperature Range	-65°C to +150°C	-65°C to +150°C
Lead Temperature (Solder, 60 sec)	300°C	300°C
Output Short Circuit Duration (Note 3)	Indefinite	Indefinite

Notes
1. Rating applies to case temperatures to 125°C; derate linearly at 6.5mW/°C for ambient temperatures above +75°C.
2. For supply voltages less than ±15V, the absolute maximum input voltage is equal to the supply voltage.
3. Short circuit may be to ground or either supply. Rating applies to +125°C case temperature or +75°C ambient temperature.

PIN CONFIGURATIONS

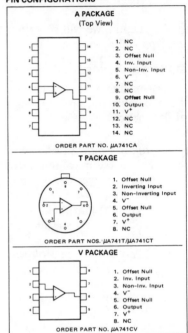

A PACKAGE
(Top View)

1. NC
2. NC
3. Offset Null
4. Inv. Input
5. Non-Inv. Input
6. V⁻
7. NC
8. NC
9. Offset Null
10. Output
11. V⁺
12. NC
13. NC
14. NC

ORDER PART NO. μA741CA

T PACKAGE

1. Offset Null
2. Inverting Input
3. Non-Inverting Input
4. V⁻
5. Offset Null
6. Output
7. V⁺
8. NC

ORDER PART NOS. μA741T/μA741CT

V PACKAGE

1. Offset Null
2. Inv. Input
3. Non-Inv. Input
4. V⁻
5. Offset Null
6. Output
7. V⁺
8. NC

ORDER PART NO. μA741CV

EQUIVALENT CIRCUIT

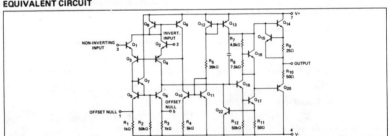

Fig. 1-2. μA741 high-performance

10

ELECTRICAL CHARACTERISTICS ($V_S = \pm 15V$, $T_A = 25°C$ unless otherwise specified)

PARAMETER	MIN.	TYP.	MAX.	UNITS	TEST CONDITIONS
μA741C					
Input Offset Voltage		2.0	6.0	mV	$R_S \leqslant 10k\Omega$
Input Offset Current		20	200	nA	
Input Bias Current		80	500	nA	
Input Resistance	0.3	2.0		MΩ	
Input Capacitance		1.4		pF	
Offset Voltage Adjustment Range		±15		mV	
Input Voltage Range	±12	±13		V	
Common Mode Rejection Ratio	70	90		dB	$R_S \leqslant 10k\Omega$
Supply Voltage Rejection Ratio		10	150	μV/V	$R_S \leqslant 10k\Omega$
Large-Signal Voltage Gain	20,000	200,000			$R_L \geqslant 2k\Omega$, $V_{out} = \pm 10V$
Output Voltage Swing	±12	±14		V	$R_L \geqslant 10k\Omega$
	±10	±13		V	$R_L \geqslant 2k\Omega$
Output Resistance		75		Ω	
Output Short-Circuit Current		25		mA	
Supply Current		1.4	2.8	mA	
Power Consumption		50	85	mW	
Transient Response (unity gain)					$V_{in} = 20mV$, $R_L = 2k\Omega$, $C_L \leqslant 100pF$
Risetime		0.3		μs	
Overshoot		5.0		%	
Slew Rate		0.5		V/μs	$R_L \geqslant 2k\Omega$
The following specifications apply for $0°C \leqslant T_A \leqslant +70°C$					
Input Offset Voltage			7.5	mV	
Input Offset Current			300	nA	
Input Bias Current			800	nA	
Large-Signal Voltage Gain	15,000				$R_L \geqslant 2k\Omega$, $V_{out} = \pm 10V$
Output Voltage Swing	±10	±13		V	$R_L \geqslant 2k\Omega$
μA741					
Input Offset Voltage		1.0	5.0	mV	$R_S \leqslant 10k\Omega$
Input Offset Current		10	200	nA	
Input Bias Current		80	500	nA	
Input Resistance	0.3	2.0		MΩ	
Input Capacitance		1.4		pF	
Offset Voltage Adjustment Range		±15		mV	
Large-Signal Voltage Gain	50,000	200,000			$R_L \geqslant 2k\Omega$, $V_{out} = \pm 10V$
Output Resistance		75		Ω	
Output Short Circuit Current		25		mA	
Supply Current		1.4	2.8	mA	
Power Consumption		50	85	mW	
Transient Response (unity gain)					$V_{in} = 20mV$, $R_L = 2k\Omega$, $C_L \leqslant 100pF$
Risetime		0.3		μs	
Overshoot		5.0		%	
Slew Rate		0.5		V/μs	$R_L \geqslant 2k\Omega$
The following specifications apply for $-55°C \leqslant T_A \leqslant +125°C$					
Input Offset Voltage		1.0	6.0	mV	$R_S \leqslant 10k\Omega$
Input Offset Current		7.0	200	nA	$T_A = +125°C$
		20	500	nA	$T_A = -55°C$
Input Bias Current		0.03	0.5	μA	$T_A = +125°C$
		0.3	1.5	μA	$T_A = -55°C$
Input Voltage Range	±12	±13		V	
Common Mode Rejection Ratio	70	90		dB	$R_S \leqslant 10k\Omega$
Supply Voltage Refection Ratio		10	150	μV/V	$R_S \leqslant 10k\Omega$
Large-Signal Voltage Gain	25,000				$R_L \geqslant 2k\Omega$, $V_{out} = \pm 10V$
Output Voltage Swing	±12	±14		V	$R_L \geqslant 10k\Omega$
	±10	±13		V	$R_L \geqslant 2k\Omega$
Supply Current		1.5	2.5	mA	$T_A = +125°C$
		2.0	3.3	mA	$T_A = -55°C$
Power Consumption		45	75	mW	$T_A = +125°C$
		45	100	mW	$T_A = -55°C$

Courtesy Signetics Corp.

operational amplifier.

TYPICAL CHARACTERISTIC CURVES

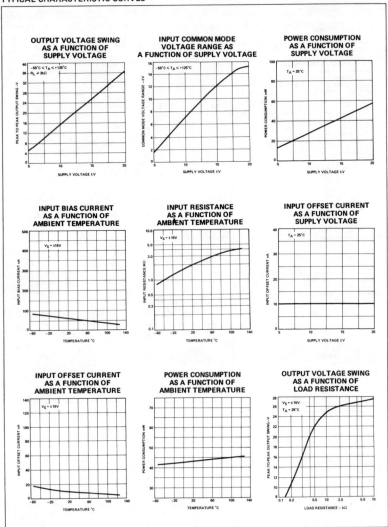

Fig. 1-2 (Cont). μA741 high-performance

TYPICAL CHARACTERISTIC CURVES (Cont'd.)

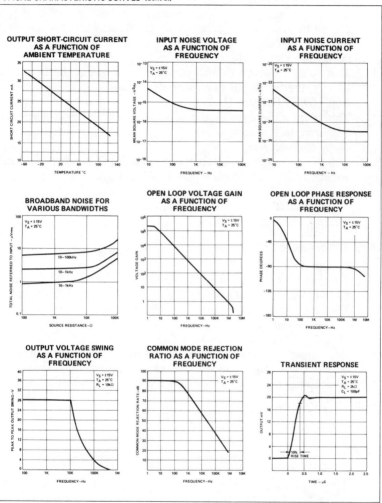

Courtesy Signetics Corp.

operational amplifier.

Input Parameters:

1. *Input Offset Voltage* (V_{oi})
   This is the voltage that must be applied to one of the input terminals to give a zero output voltage. Remember, for an *ideal* op-amp, the output voltage offset is zero!

2. *Input Bias Current* (I_b)
   This is the average of the currents flowing into both inputs. Ideally, the two input bias currents are equal.

3. *Input Offset Current* (I_{os})
   This is the difference of the two input bias currents when the output voltage is zero.

4. *Input Voltage Range* (V_{cm})
   This is the range of the common-mode input voltage (i.e., the voltage common to both inputs and ground).

5. *Input Resistance* (Z_i)
   This is the resistance "looking in" at either input with the remaining input grounded.

Output Parameters:

1. *Output Resistance* (Z_{oi})
   This is the resistance seen "looking into" the op-amp's output.

2. *Output Short-Circuit Current* (I_{osc})
   This is the maximum output current that the op-amp can deliver to a load.

3. *Output Voltage Swing* ($\pm V_o$ max)
   Depending on the load resistance, this is the maximum *peak* output voltage that the op-amp can supply without saturation or clipping.

Dynamic Parameters:

1. *Open-Loop Voltage Gain* (A_{OL})
   This is the ratio of the output to input voltage of the op-amp *without external feedback*.

2. Large-Signal Voltage Gain
   This is the ratio of the maximum voltage swing to the change in the input voltage required to drive the output from zero to a specified voltage (e.g., ± 10 volts).

3. *Slew Rate* (SR)
   This is the time rate of change of the output voltage with the op-amp circuit having a voltage gain of unity (1.0).

Other Parameters:

1. *Supply Current*
   This is the current that the op-amp will draw from the power supply.

2. *Common-Mode Rejection Ratio* (CMRR)
   This is a measure of the ability of the op-amp to reject signals that are simultaneously present at both inputs. It is the ratio of the common-mode input voltage to the generated output voltage, usually expressed in decibels (dB).

3. *Channel Separation*
   Whenever there is more than one op-amp in a single package, such as a type 747 op-amp, a certain amount of "crosstalk" will be present. That is, a signal applied to the input of one section of a dual op-amp will produce a finite output signal in the remaining section, *even though there is no input signal applied to the unused section.*

Don't worry if the implication of most of these parameters is not clear to you now. As you read through this book and perform the experiments, you will understand their significance.

GAIN AND FREQUENCY RESPONSE

Unlike the ideal op-amp, the op-amp that is used in various circuits does not have infinite gain and bandwidth. As shown in Fig. 1-3, the open-loop gain A_{OL} for a type 741 op-amp is graphed as a function of frequency. At very low frequencies, the open-loop gain of an op-amp is constant, but begins to "roll off" at approximately 6 Hz at a rate of -6 dB/octave or -20 dB/decade. *An octave is a doubling in frequency, and a decade is a ten-fold increase in frequency.* This decrease continues until the gain is unity, or 0 dB. The frequency at which the gain is unity is called the unity gain frequency, f_T.

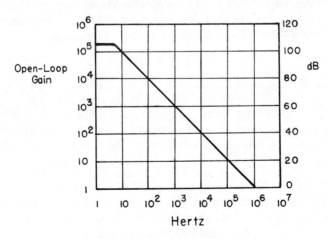

Fig. 1-3. 741 op-amp open-loop gain.

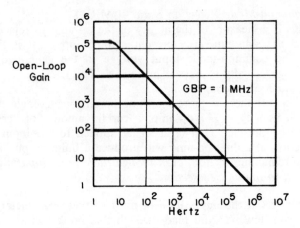

Fig. 1-4. Bandwidth for the 741 op-amp.

As we shall see in Chapter 2, when some of the output signal is fed back to the op-amp's input, the ratio of the output to input voltage is termed the *closed-loop gain,* A_{CL}, and is always less than the open-loop gain. The difference in decibels between the open-loop and closed-loop gains is the *loop gain,* A_L. When A_{OL} and A_{CL} are expressed as simple output-to-input ratios, the loop gain is expressed mathematically as:

$$A_L = \frac{A_{OL}}{A_{CL}} \qquad \text{(Eq. 1-1)}$$

Perhaps the first factor in the consideration of a particular op-amp for a given application is its *gain-bandwidth product,* or GBP. For the response curve of Fig. 1-3, *the product of the open-loop gain and frequency is a constant at any point on the curve,* so that:

$$GBP = A_{OL}BW \qquad \text{(Eq. 1-2)}$$

Graphically, the bandwidth is the point at which the closed-loop gain curve intersects the open-loop gain curve, as shown in Fig. 1-4 for a family of closed-loop gains. Therefore, one obtains the bandwidth for any desired closed-loop gain by simply drawing a horizontal line from the desired value of gain to intersect the rolloff of the open-loop gain curve.

For a practical design situation, the actual design gain of an op-amp circuit should be about a factor of 1/10 to 1/20 of the open-loop gain at a given frequency. This ensures that the op-amp will function properly without distortion. As an example, using the response of Fig. 1-3, the closed-loop gain at 10 kHz should be about 5 to 10, since the open-loop gain is 100 (40 dB).

Before we leave this section, one additional parameter is worth mentioning. The *transient response, or rise time,* is the time that it takes for the output signal to go from 10% to 90% of its final value when a step-function pulse is used as an input signal, and is specified under closed-loop conditions. From electronic circuit theory, the rise time is related to the bandwidth of the op-amp by the relation:

$$BW = \frac{0.35}{\text{rise time}} \qquad \text{(Eq. 1-3)}$$

THE POWER SUPPLY

In general, op-amps are designed to be powered from a dual, or bipolar, voltage supply which is typically in the range of ±5 to ±15 volts. That is, one supply is +5 to +15 volts *with respect to ground,* and another supply voltage of −5 to −15 volts with respect to ground, as shown in Fig. 1-5. However, in certain cases, an op-amp may be operated from a single supply voltage, which is explained in Chapters 8 and 9.

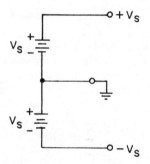

Fig. 1-5. Dual-polarity power supply.

PERFORMING THE EXPERIMENTS

In this user-oriented text/workbook, you will have the opportunity to set up and analyze experiments that demonstrate principles, concepts, and applications of many of the basic circuit configurations using op-amps. There are perhaps more experiments than you need to gain a sound understanding about the design and operation of op-amp circuits. In most cases, it is not necessary to perform every experiment in each chapter.

BREADBOARDING

The breadboard is designed to accommodate the many experiments that you will perform in this and subsequent chapters. Op-amp integrated circuit chips, resistors, capacitors, and other components as well as electrical power all connect or tie directly to the breadboard.

Fig. 1-6 shows the top view of the basic component of a typical breadboard system, which is known as the *SK-10 Universal Breadboarding Socket,* and is manufactured by E&L Instruments, Inc. It contains 64 by 2 sets of 5 electrically connected solderless terminals

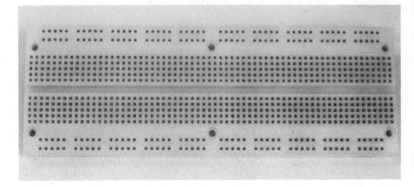

Courtesy E & L Instruments, Inc.

Fig. 1-6. Top view of the SK-10 Universal Breadboarding Socket.

that straddle both sides of a narrow center groove, and 8 sets of 25 electrically connected terminals along the edges. The center group of 5 electrically connected terminals accommodate the integrated circuit chips and permit 4 additional connections to be made at each pin of the integrated circuit chip. Another breadboarding aid, also manufactured by E&L Instruments, is shown in Fig. 1-7. Called the Model OA-2 *Op-Amp Designer,* it contains an SK-10 socket, function generator, and ±15/+5 volt power supplies. Its features and operation for use with the experiments in this book are explained in the Appendix.

RULES FOR SETTING UP THE EXPERIMENTS

Throughout this book you will be breadboarding various circuits, either using SK-10 socket, one of the op-amp designers, or some other method comfortable for you. If you already have had experience with one of the following books,

1. Larsen, D. G., and P. R. Rony. *Logic & Memory Experiments Using TTL Integrated Circuits,* Books 1 and 2. Howard W. Sams & Co., Inc.
2. Rony, P. R., Larsen, D. G., and J. A. Titus. *The 8080A Bugbook.* Howard W. Sams & Co., Inc.

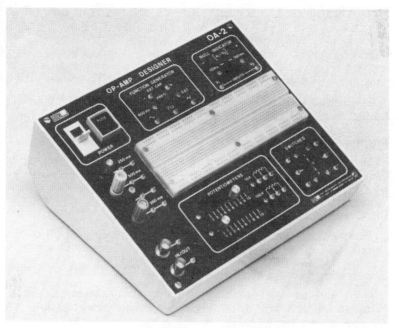

Courtesy E & L Instruments, Inc.

Fig. 1-7. Model OA-2 Op-Amp Designer.

3. Rony, P. R., Larsen, D. G., and J. A. Titus. *Introductory Experiments in Digital Electronics and 8080A Microcomputer Programming and Interfacing,* Books 1 and 2. Howard W. Sams & Co., Inc.
4. Berlin, H. M. *555 Timer Applications Sourcebook, With Experiments.* Howard W. Sams & Co., Inc.
5. Berlin, H. M. *Design of Active Filters, With Experiments.* Howard W. Sams & Co., Inc.

these rules will be familiar. Before you set up any experiment, it is recommended that you do the following:

1. Plan your experiment beforehand. Know what types of results you are expected to observe.
2. Disconnect or turn off *all* power and signal sources to the breadboard.
3. Clear the breadboard of all wires and components from previous experiments, unless stated otherwise.
4. Check the wired-up circuit against the schematic circuit diagram to make sure that it is correct.
5. Connect or turn on the power and signal sources to the breadboard *last!*

6. When finished, make sure that you disconnect or turn off all power and signal sources to the breadboard *before* you clear the breadboard of wires and components.

FORMAT FOR THE EXPERIMENTS

The instructions for each experiment are presented in the following format:

Purpose

The material presented under this heading states the purpose of your performing the experiment. It is well for you to have this intended purpose in mind as you conduct the experiment.

Pin Configuration of Integrated Circuit Chips

The pin configurations are given under this heading for all the integrated circuit chips used in the experiment.

Schematic Diagram of Circuit

Under this heading is the schematic diagram of the completed circuit that you will wire in the experiment. You should analyze this diagram in an effort to obtain an understanding of the circuit *before* you proceed further.

Design Basics

Under this heading is a brief summary of the equations used for the design of the circuit.

Steps

A series of sequential steps describe the instructions for setting up portions of the experiment. Questions are also included at different points of the section. The numerical calculations are easily accomplished using one of the many "pocket-type" calculators.

HELPFUL HINTS AND SUGGESTIONS

Tools

Only two tools are necessary for all of the experiments given in this book:

- a pair of "long-nosed" pliers.
- a wire stripper/cutter.

The pliers are used to:

- straighten out the bent ends of hook-up wire that is used to wire the circuits on the breadboard.

- straighten out or bend the resistor, capacitor, and other component leads to the proper position so that they can be conveniently inserted into the breadboard.

The wire stripper/cutter is used to cut the hook-up wire to size and strip about $\frac{3}{8}''$ of insulation from each end.

Wire

Only #22, #24, or #26 insulated wire is used, and it must be solid, not stranded!

Breadboarding

- Never insert too large a wire or component lead into a breadboard terminal.

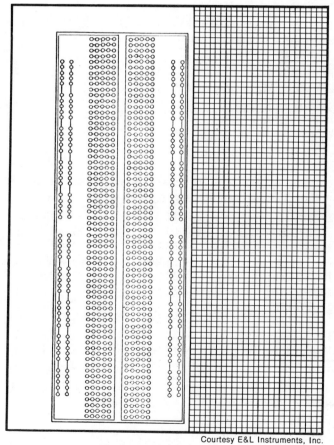

Courtesy E&L Instruments, Inc.

Fig. 1-8. SK-10 layout sheet.

- Never insert a bent wire. Straighten out the end with a pair of pliers before insertion.
- Try to maintain an orderly arrangement of components and wires, keeping all connections as short as possible.
- Plan the construction of your circuit on a layout sheet, like the one shown in Fig. 1-8, before you breadboard it. Layout sheets can be obtained in quantity from E&L Instruments, Inc.

Components and Other Equipment

A variety of resistors, capacitors, etc. will be required for the experiments. For those of you who regularly experiment with electronics, your "junkbox" should yield most of these. The majority of the following components can be obtained from a number of suppliers that cater to experimenters, some of which are listed below:

- James Electronics
  1021-A Howard Avenue
  San Carlos, California 94070

- Poly Paks
  P. O. Box 942
  Lynnfield, Massachusetts 01940

- Solid State Sales
  P. O. Box 74A
  Somerville, Massachusetts 02143

- *Resistors*

  The following fixed resistors needed to perform all the experiments should be 5%, and can be either $\frac{1}{4}$ or $\frac{1}{2}$ watt:

2—100 Ω	1—8.2 kΩ	3—100 kΩ
1—270 Ω	3—10 kΩ	1—150 kΩ
1—470 Ω	1—12 kΩ	1—180 kΩ
3—1 kΩ	2—15 kΩ	1—200 kΩ
1—1.5 kΩ	2—20 kΩ	1—220 kΩ
1—2.2 kΩ	1—22 kΩ	1—270 kΩ
1—2.7 kΩ	1—27 kΩ	2—330 kΩ
1—3.3 kΩ	3—33 kΩ	1—390 kΩ
1—3.9 kΩ	1—39 kΩ	1—470 kΩ
1—4.7 kΩ	2—47 kΩ	2—1.0 MΩ
1—5.6 kΩ	2—68 kΩ	1—1.5 MΩ
1—6.8 kΩ	1—82 kΩ	

  The following potentiometers are also required:

  1—50 kΩ, single turn
  1—100 kΩ, 10-turn
  1—5 MΩ, single turn

- *Capacitors*

  The following capacitors are required:

  1—.001 μF
  2—.0022 μF
  1—.0047 μF
  2—.01 μF
  3—.033 μF
  1—.047 μF
  1—.1 μF
  1—5 μF

- *Other Components*

  1—general-purpose photocell
  1—light-emitting diode (LED)
  3—1N914 diode
  1—1N751, 5.1-volt zener diode
  2—1N746, 3.3-volt zener diode
  1—2N2222 npn transistor
  1—4011 CMOS quad 2-input NAND gate
  1—4016 CMOS analog switch
  1—spdt switch

- *Oscilloscope*—Just about any general-purpose oscilloscope will do. A single-trace model is OK, but a dual-trace unit is preferred, as it will be useful for rapid comparison of the input and output signals.
- *VOM, VTVM, or Digital Multimeter*—A general-purpose meter that is capable of measuring ac and dc voltage and current, and resistance is necessary. Please use a digital-type meter if you can get one, as the resolution of the measurements will be greatly improved.
- *Function Generator*—A function generator capable of producing sine, square, and triangle outputs with adjustable frequency and amplitude. For sine-wave output, the following schematic symbol will be used:

For square waves, the following symbol will be used:

And in a similar fashion, the following symbol will be used for triangle waves:

- *Pocket Calculator*—This is not mandatory, but it is recommended that you use one. The routine calculations can be accomplished with the simplest of the 4-function calculators. It sure saves a lot of time, especially when you can now get one for less than $15.

Fig. 1-9 illustrates the types of equipment and hardware that you will generally need for the experiments in this book.

Op-Amp Chips

Although there are many types of op-amps available, only one or two types will be used for just about all of the experiments in this book. These are of the "low-cost" variety that are very popular with hobbyist and scientists with limited funds.

For your experiments, *use only the op-amps that are dual-in-line (DIP)chips, as these are the only ones that are conveniently used with the solderless breadboarding sockets.* These chips are similar in appearance to the 14- and 16-pin TTL and CMOS integrated circuit chips.

For almost all of the experiments, we will use the type 741 op-amp. It is perhaps the most widely used op-amp, costing approximately 35 cents each. Although the 741 op-amp comes in several different package styles, the 8-pin "mini-DIP" (sometimes called the "V" package) is preferred.

To perform all the experiments in this book, you will need the following op-amp chips:

- 3—741 op-amps
- LM318 op-amp
- LM3900 op-amp
- 3660J Burr-Brown instrumentation amplifier

The following pages present the data sheets for the following popular op-amps, some of which you may want to use after you finish this book.

- μA747 (Signetics)
- 5558 (Signetics)
- LM118/218/318 (National Semiconductor)
- 4156 (Raytheon)

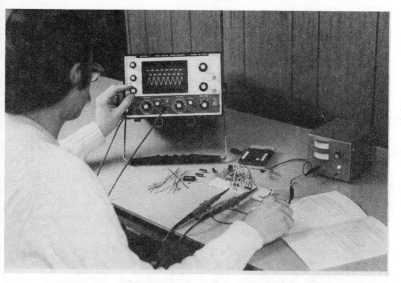

Fig. 1-9. Equipment and hardware used for experiments in this book.

AN INTRODUCTION TO THE EXPERIMENTS

The following experiments are concerned with the measurement of several of the characteristic parameters associated with the 741C op-amp. This particular op-amp, in addition to being cheap (about 35 cents each), is highly representative of the general-purpose op-amps that are available. With the exception of Chapters 9 and 10, all the experiments that you will perform will use the 741C op-amp. Consequently, you will become familiar with its characteristics.

The experiments that you will perform can be summarized as follows:

Experiment No.	Purpose
1	Measure the input offset voltage of an op-amp.
2	Measure the input bias currents of an op-amp.
3	Measure the input impedance of an op-amp.
4	Measure the slew rate of an op-amp.
5	Measure the common-mode rejection ratio (CMRR) of an op-amp.
6	Determine the closed-loop response of an op-amp by calculating its gain-bandwidth product.

DUAL OPERATIONAL AMPLIFIER µA747

LINEAR INTEGRATED CIRCUITS

DESCRIPTION

The µA747 is a pair of high performance monolithic operational amplifiers constructed on a single silicon chip. They are intended for a wide range of analog applications where board space or weight are important. High common mode voltage range and absence of "latch-up" make the µA747 ideal for use as a voltage follower. The high gain and wide range of operating voltage provides superior performance in integrator, summing amplifier, and general feedback applications. The µA747 is short-circuit protected and requires no external components for frequency compensation. The internal 6 db/octave roll-off insures stability in closed loop applications. For single amplifier performance, see µA741 data sheet.

FEATURES

- NO FREQUENCY COMPENSATION REQUIRED
- SHORT-CIRCUIT PROTECTION
- OFFSET VOLTAGE NULL CAPABILITY
- LARGE COMMON-MODE AND DIFFERENTIAL
- VOLTAGE RANGES
- LOW POWER CONSUMPTION
- NO LATCH UP

ABSOLUTE MAXIMUM RATINGS

Supply Voltage µA747		+22V
µA747C		+18V
Internal Power Dissipation (Note 1)	Metal Can	500 mW
	DIP	670 mW
Differential Input Voltage		+30V
Input Voltage (Note 2)		+15V
Voltage between Offset Null and V^-		+0.5V
Storage Temperature Range		-65°C to +155°C
Operating Temperature Range µA747		-55°C to +125°C
µA747C		0°C to +70°C
Lead Temperature (Soldering 60 seconds)		300°C
Output Short Circuit Duration (Note 3)		Indefinite

PIN CONFIGURATION

A PACKAGE
(Top View)

1. Inv Input A
2. Non-inv Input A
3. Offset Null A
4. V^-
5. Offset Null B
6. Non-inv Input B
7. Inv Input B
8. Offset Null B
9. V+ B
10. Output B
11. No Connect
12. Output A
13. V+ A
14. Offset Null A

ORDER PART NOS.
µA747A
µA747CA

K PACKAGE

1. Output A
2. V^+ A
3. Inverting Input A
4. Non-inverting Input A
5. V^-
6. Non-inverting Input B
7. Inverting Input B
8. V^+ B
9. Output B
10. NC

ORDER PART NOS.
µA747K
µA747CK

EQUIVALENT CIRCUIT (Each Side)

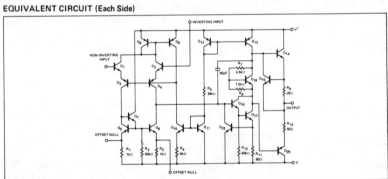

Fig. 1-10. µA747 dual

ELECTRICAL CHARACTERISTICS ($V_S = \pm 15$ V, $T_A = 25^{\circ}$C unless otherwise specified)

PARAMETERS	CONDITIONS	MIN.	TYP.	MAX.	UNITS
Input Offset Voltage	$R_S \leq 10$ kΩ		1.0		mV
μA747			5.0		mV
μA747C			6.0		mV
Input Offset Current			20	200	nA
Input Bias Current			80	500	nA
Input Resistance		0.3	2.0		MΩ
Input Capacitance			1.4		pF
Offset Voltage Adjustment Range			± 15		mV
Large-Signal Voltage Gain	$R_L \geq 2K\Omega, V_{out} = \pm 10$V		200,000		
μA747		50,000			
μA747C		25,000			
Output Resistance			75		Ω
Output Short-Circuit Current			25		mA
Supply Current			1.7	2.8	mA
Power Consumption			50	85	mW
Transient Response (unity gain)	$V_{in} = 20$ mV, $R_L = 2k\Omega$, $C_L \leq 100$ pF				
Risetime			0.3		μS
Overshoot			5.0		%
Slew Rate	$R_L \geq 2k\Omega$		0.5		V/μS
Channel Separation			120		dB

μA747

The following specifications apply for -55°C $\leq T_A \leq +125^{\circ}$C

PARAMETERS	CONDITIONS	MIN.	TYP.	MAX.	UNITS
Input Offset Voltage	$R_S \leq 10$kΩ		1.0	6.0	mV
Input Offset Current	$T_A = +125^{\circ}$C		7.0	200	nA
	$T_A = -55^{\circ}$C		85	500	nA
Input Bias Current	$T_A = +125^{\circ}$C		0.03	0.5	μA
	$T_A = -55^{\circ}$C		0.3	1.5	μA
Input Voltage Range		± 12	± 13		V
Common Mode Rejection Ratio	$R_S \leq 10$kΩ	70	90		dB
Supply Voltage Rejection Ratio	$R_S \leq 10$kΩ		30	150	μV/V
Large-Signal Voltage Gain	$R_L \geq 2k\Omega, V_{out} = \pm 10$V	25,000			
Output Voltage Swing	$R_L \geq 10$kΩ	± 12	± 14		V
	$R_L \geq 2k\Omega$	± 10	± 13		V
Supply Current	$T_A = +125^{\circ}$C		1.5	2.5	mA
	$T_A = -55^{\circ}$C		2.0	3.3	mA
Power Consumption	$T_A = +125^{\circ}$C		45	75	mW
	$T_A = -55^{\circ}$C		60	100	mW

μA747C

The following specifications apply for 0°C $\leq T_A \leq +70^{\circ}$C

PARAMETERS	CONDITIONS	MIN.	TYP.	MAX.	UNITS
Input Offset Voltage	$R_S \leq 10$ kΩ		1.0	7.5	mV
Input Offset Current			7.0	300	nA
Input Bias Current			0.03	0.8	μA
Input Voltage Range		± 12	± 13		V
Common Mode Rejection Ratio	$R_S \leq 10$ kΩ	70	90		dB
Supply Voltage Rejection Ratio	$R_S \leq 10$ kΩ		30	150	μV/V
Large-Signal Voltage Gain	$R_L \geq 2k\Omega, V_{out} = \pm 10$V	15,000			
Output Voltage Swing	$R_L \geq 10$kΩ	± 12	± 14		V
	$R_L \geq 2k\Omega$	± 10	± 13		V
Supply Current			2.0	3.3	mA
Power Consumption			60	100	mW

NOTES:

1. Rating applied to ambient temperatures up to 70°C ambient derate linearly at 6.3 mW/$^{\circ}$C for the Metal Can and 8.3 mW/$^{\circ}$C for the Ceramic DIP package.

2. For supply voltages less than +15V, the absolute maximum input voltage is equal to the supply voltage.

3. Short circuit may be to ground or either supply. Military rating applies to +125°C case temperature or +60°C ambient temperature for each side.

Courtesy Signetics Corp.

operational amplifier.

Signetics

DESCRIPTION

The 5558 consists of a pair of high performance monolithic operational amplifiers constructed on a single chip. It features internal compensation and is intended for use in a variety of analog applications. High common mode voltage range and immunity to latch-up makes the 5558 ideal for use as a voltage follower. The high gain and wide range of operating voltage achieves superior performance in integrator, summing amplifier, and general feedback applications. The device is short-circuit protected. For single amplifier performance see the 5741 data sheet. The 5558 is a pin-for-pin replacement for the MC1558G.

ABSOLUTE MAXIMUM RATINGS

Power Supply Voltages
S5558	±22V
N5558	±18V
Differential Input Voltage	±30V
Common-mode Input Swing	±15V
Output Short Circuit Duration	Continuous

Power Dissipation (Note 1)
T Package — (MO-002-AG)	680mW
V Package	625mW

Operating Temperature Range
S5558	–55°C to +125°C
N5558	0°C to +75°C
Storage Temperature Range	–65°C to +150°C
Lead Temperature (Soldering, 60 sec)	300°C

NOTE:
1. Derate T package linearly at 4.6 mW/°C for ambient temperatures above +25°C
2. Derate V package at 5mW/°C above 25°C

PIN CONFIGURATIONS

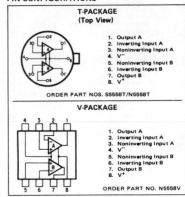

T-PACKAGE (Top View)

1. Output A
2. Inverting Input A
3. Noninverting Input A
4. V⁻
5. Noninverting Input B
6. Inverting Input B
7. Output B
8. V⁺

ORDER PART NOS. S5558T/N5558T

V-PACKAGE

1. Output A
2. Inverting Input A
3. Noninverting Input A
4. V⁻
5. Noninverting Input B
6. Inverting Input B
7. Output B
8. V⁺

ORDER PART NO. N5558V

FEATURES:

- 2 "OP AMPS" IN SPACE OF ONE 741 V PACKAGE
- NO FREQUENCY COMPENSATION REQUIRED
- SHORT CIRCUIT PROTECTION
- LOW POWER CONSUMPTION
- LARGE COMMON MODE AND DIFFERENTIAL VOLTAGE RANGES
- NO LATCH-UP

EQUIVALENT SCHEMATIC

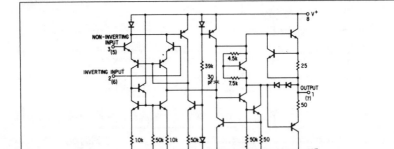

The numbers without parenthesis represent the pin numbers for ½ of the dual circuit. The numbers in parenthesis represent the pin numbers for the other half.

Fig. 1-11. 5558 dual

ELECTRICAL CHARACTERISTICS (V^+ = +15 Vdc, V^- = –15 Vdc, T_A = +25°C unless otherwise noted)

CHARACTERISTICS	SYMBOL	MIN		TYP		MAX		UNIT
		S5558	N5558	S5558	N5558	S5558	N5558	
Input Bias Current T_A = +25°C	I_b			0.2	0.2	0.5	0.5	µAdc
T_A = T_{low} to T_{high} (See Note 1)						1.5	0.8	
Input Offset Current T_A = +25°C	$|I_{io}|$			0.03	0.03	0.2	0.2	µAdc
T_A = T_{low} to T_{high}						0.5	0.3	
Input Offset Voltage ($R_S \leqslant 10k\Omega$) T_A = +25°C	$|V_{io}|$			1.0	2.0	5.0	6.0	mVdc
T_A = T_{low} to T_{high}						6.0	7.5	
Differential Input Impedance (Open-Loop, f = 20 Hz) Parallel Input Resistance	R_p	0.3	0.3	1.0	1.0			Megohm
Parallel Input Capacitance	C_p			6.0	6.0			pF
Common-Mode Input Impedance (f = 20 Hz)	$Z_{(in)}$			200	200			Megohms
Common-Mode Input Voltage Swing	CMV_{in}	±12	±12	±13	±13			V_{pk}
Equivalent Input Noise Voltage (A_V = 100, R_s = kΩ, f = 1.0 kHz, BW = 1.0 Hz)	e_n			45	45			nV(Hz)½
Common-Mode Rejection Ratio (f = 100 Hz)	CM_{rej}	70	70	90	90			dB
Open-Loop Voltage Gain, (V_{out} = ±10V, R_L = 2.0kΩ)	A_{VOL}							V/V
T_A = +25°C		50,000	20,000	200,000	100,000			
T_A = T_{low} to T_{high}		25,000	15,000					
Power Bandwidth (A_V = 1, R_L = 2.0kΩ, THD≤5%,V_{out} = 20Vp-p)	P_{BW}			14	14			kHz
Unity Gain Crossover Frequency (open-loop)				1.1	1.1			MHz
Phase Margin (open-loop, unity gain)				65	65			degrees
Gain Margin				11	11			dB
Slew Rate (Unity Gain)	dV_{out}/dt			0.8	0.8			V/µs
Output Impedance (f = 20 Hz)	Z_{out}			300	300			ohms
Short-Circuit Output Current	I_{SC}			20	20			mAdc
Output Voltage Swing (R_L = 10kΩ)	V_{out}	±12	±12	±14	±14			V_{pk}
R_L = 2kΩ (T_A = T_{low} to T_{high})		±10	±10	±13	±13			
Power Supply Sensitivity V^- = constant, $R_s \leqslant 10k\Omega$	S^+			30	30	150	150	µV/V
V^+ = constant, $R_s \leqslant 10k\Omega$	S^-			30	30	150	150	
Power Supply Current	$I_D{}^+$			2.3	2.3	5.0	5.6	mAdc
	$I_D{}^-$			2.3	2.3	5.0	5.6	
DC Quiescent Power Dissipation (V_{out} = 0)	P_D			70	70	150	170	mW
Channel Separation	e_{o1}/e_{o2}			120	120			dB

Note 1: T_{low}: 0°C for N5558, –55°C for S5558; T_{high}: +75°C for N5558, +125°C for S5558

TYPICAL CHARACTERISTIC CURVES

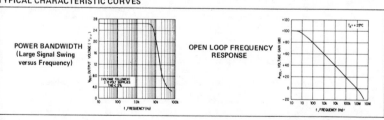

POWER BANDWIDTH (Large Signal Swing versus Frequency)

OPEN LOOP FREQUENCY RESPONSE

Courtesy Signetics Corp.

operational amplifiers.

Operational Amplifiers/Buffers

LM118/LM218/LM318 operational amplifier

general description

The LM118 series are precision high speed operational amplifiers designed for applications requiring wide bandwidth and high slew rate. They feature a factor of ten increase in speed over general purpose devices without sacrificing DC performance.

features

- 15 MHz small signal bandwidth
- Guaranteed 50V/μs slew rate
- Maximum bias current of 250 nA
- Operates from supplies of ±5V to ±20V
- Internal frequency compensation
- Input and output overload protected
- Pin compatible with general purpose op amps

The LM118 series has internal unity gain frequency compensation. This considerably simplifies its application since no external components are necessary for operation. However, unlike most internally compensated amplifiers, external frequency compensation may be added for optimum performance For inverting applications, feedforward compensation will boost the slew rate to over 150V/μs and almost double the bandwidth. Overcompensation can be used with the amplifier for greater stability when maximum bandwidth is not needed. Further, a single capacitor can be added to reduce the 0.1% settling time to under 1 μs.

The high speed and fast settling time of these op amps make them useful in A/D converters, oscillators, active filters, sample and hold circuits, or general purpose amplifiers. These devices are easy to apply and offer an order of magnitude better AC performance than industry standards such as the LM709.

The LM218 is identical to the LM118 except that the LM218 has its performance specified over a −25°C to +85°C temperature range. The LM318 is specified from 0°C to +70°C.

schematic and connection diagrams

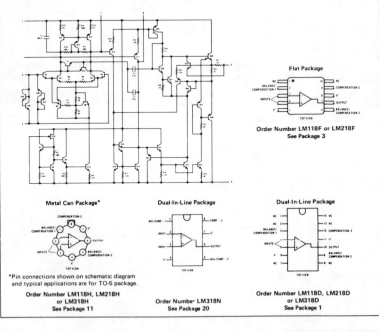

Flat Package

Order Number LM118F or LM218F
See Package 3

Metal Can Package*

*Pin connections shown on schematic diagram and typical applications are for TO-5 package.

Order Number LM118H, LM218H
or LM318H
See Package 11

Dual-In-Line Package

Order Number LM318N
See Package 20

Dual-In-Line Package

Order Number LM118D, LM218D
or LM318D
See Package 1

Fig. 1-12. LM118/LM218/LM318

absolute maximum ratings

Supply Voltage	±20V
Power Dissipation (Note 1)	500 mW
Differential Input Current (Note 2)	±10 mA
Input Voltage (Note 3)	±15V
Output Short-Circuit Duration	Indefinite
Operating Temperature Range	
LM118	−55°C to +125°C
LM218	−25°C to +85°C
LM318	0°C to +70°C
Storage Temperature Range	−65°C to +150°C
Lead Temperature (Soldering, 10 seconds)	300°C

electrical characteristics (Note 4)

PARAMETER	CONDITIONS	LM118/LM218			LM318			UNITS
		MIN	TYP	MAX	MIN	TYP	MAX	
Input Offset Voltage	$T_A = 25°C$		2	4		4	10	mV
Input Offset Current	$T_A = 25°C$		6	50		30	200	nA
Input Bias Current	$T_A = 25°C$		120	250		150	500	nA
Input Resistance	$T_A = 25°C$	1	3		0.5	3		MΩ
Supply Current	$T_A = 25°C$		5	8		5	10	mA
Large Signal Voltage Gain	$T_A = 25°C, V_S = ±15V$ $V_{OUT} = ±10V, R_L \geq 2 k\Omega$	50	200		25	200		V/mV
Slew Rate	$T_A = 25°C, V_S = ±15V, A_v = 1$	50	70		50	70		V/µs
Small Signal Bandwidth	$T_A = 25°C, V_S = ±15V$		15			15		MHz
Input Offset Voltage				6			15	mV
Input Offset Current				100			300	nA
Input Bias Current				500			750	nA
Supply Current	$T_A = 125°C$		4.5	7				mA
Large Signal Voltage Gain	$V_S = ±15V, V_{OUT} = ±10V$ $R_L \geq 2 k\Omega$	25			20			V/mV
Output Voltage Swing	$V_S = ±15V, R_L = 2 k\Omega$	±12	±13		±12	±13		V
Input Voltage Range	$V_S = ±15V$	±11.5			±11.5			V
Common-Mode Rejection Ratio		80	100		70	100		dB
Supply Voltage Rejection Ratio		70	80		65	80		dB

Note 1: The maximum junction temperature of the LM118 is 150°C, the LM218 is 100°C, and the LM318 is 85°C. For operating at elevated temperatures, devices in the TO-5 package must be derated based on a thermal resistance of 150°C/W, junction to ambient, or 45°C/W, junction to case. For the flat package, the derating is based on a thermal resistance of 185°C/W when mounted on a 1/16-inch-thick epoxy glass board with ten, 0.03-inch-wide, 2-ounce copper conductors. The thermal resistance of the dual-in-line package is 100°C/W, junction to ambient.

Note 2: The inputs are shunted with back-to-back diodes for overvoltage protection. Therefore, excessive current will flow if a differential input voltage in excess of 1V is applied between the inputs unless some limiting resistance is used.

Note 3: For supply voltages less than ±15V, the absolute maximum input voltage is equal to the supply voltage.

Note 4: These specifications apply for ±5V $\leq V_S \leq$ ±20V and −55°C $\leq T_A \leq$ +125°C, (LM118), −25°C $\leq T_A \leq$ +85°C (LM218), and 0°C $\leq T_A \leq$ +70°C (LM318). Also, power supplies must be bypassed with 0.1µF disc capacitors.

Courtesy National Semiconductor Corp.

operational amplifier.

typical performance characteristics LM118, LM218

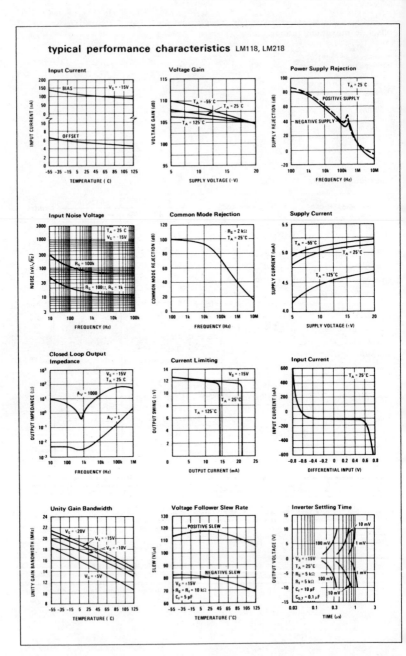

Fig. 1-12 (Cont.). LM118/LM218/LM318

32

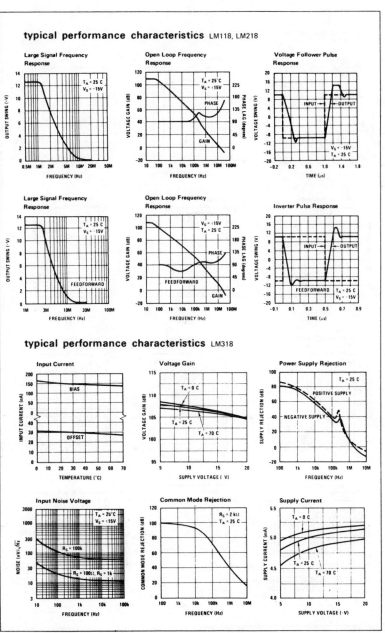

typical performance characteristics LM118, LM218

typical performance characteristics LM318

Courtesy National Semiconductor Corp.

operational amplifier.

typical performance characteristics LM318 (Cont'd)

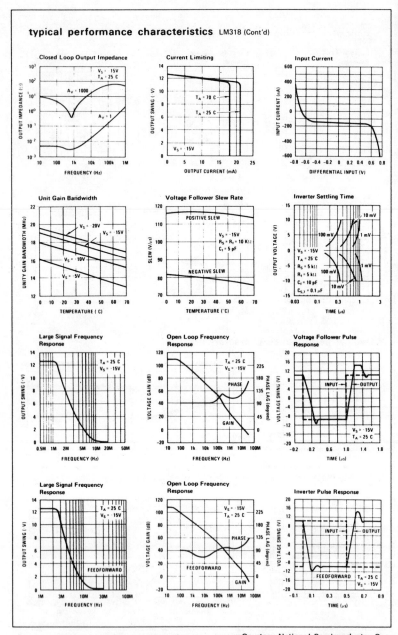

Courtesy National Semiconductor Corp.

Fig. 1-12 (Cont.). LM118/LM218/LM318 operational amplifier.

QUAD
HIGH PERFORMANCE
OPERATIONAL AMPLIFIER
4156

DESCRIPTION

The RM4156/RC4156 is a monolithic integrated circuit, consisting of four independent high performance operational amplifiers constructed with the planar epitaxial process.

These amplifiers feature guaranteed A.C. performance which far exceeds that of the 741 type amplifiers. Also featured are excellent input characteristics and guaranteed low noise making this device the optimum choice for audio, active filter and instrumentation applications.

FEATURES

	Typical	Guaranteed
● Unity Gain Bandwidth	3.5 MHz	2.8 MHz
● High Slew Rate	1.6V/μS	1.3V/μS
● Low Noise Voltage	0.8μV	2.0μV

- ● Indefinite Short Circuit Protection
- ● No Crossover Distortion
- ● Low Input Offset and Bias Parameters
- ● Internal Compensation

SCHEMATIC DIAGRAM

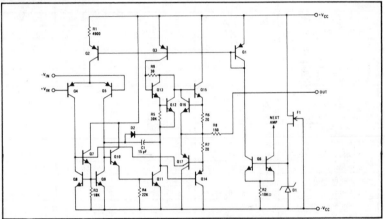

CONNECTION INFORMATION

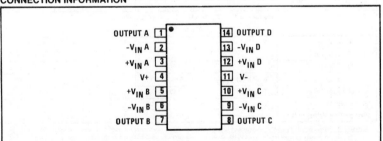

RAYTHEON COMPANY ● Semiconductor Division ● 350 Ellis Street ● Mountain View, CA 94042
Printed in U.S.A.   SEPTEMBER 1976

Courtesy Raytheon Co.

Fig. 1-13. 4156 quad high-performance operational amplifier.

ABSOLUTE MAXIMUM RATINGS

Supply Voltage	±20V	Storage Temperature Range	−65 to +150°C
Internal Power Dissipation (Note 1)	880 mW	Operating Temperature Range RM4156	−55 to +125°C
Differential Input Voltage	±30V	RV4156	−40 to +85°C
Input Voltage (Note 2)	±15V	RC4156	0 to +70°C
Output Short Circuit Duration (Note 3)	Indefinite	Lead Soldering Temperature (60 sec)	300°C

ELECTRICAL CHARACTERISTICS V_{CC} ±15V T_A +25°C unless otherwise specified

PARAMETER	CONDITIONS	RM4156			RV4156/RC4156			UNITS
		MIN	TYP	MAX	MIN	TYP	MAX	
Input Offset Voltage	$R_S \leqslant 10\ K\Omega$		0.5	3.0		1.0	5.0	mV
Input Offset Current			15	30		30	50	nA
Input Bias Current			60	200		60	300	nA
Input Resistance			5			5		MΩ
Large Signal Voltage Gain	$R_L \geqslant 2\ K\Omega\ V_{OUT}$ ±10V	50,000	100,000		25,000	100,000		V/V
Output Voltage Swing	$R_L \geqslant 10\ K\Omega$	±12	±14		±12	±14		V
	$R_L \geqslant 2\ K\Omega$	±10	±13		±10	±13		V
Input Voltage Range		±12	±14		±12	±14		V
Output Resistance			230			230		Ω
Output Short Circuit Current			25			25		mA
Common Mode Rejection Ratio	$R_S \leqslant 10\ K\Omega$	80			80			dB
Power Supply Rejection Ratio	$R_S \leqslant 10\ K\Omega$	80			80			dB
Supply Current (all amplifiers)	$R_L = \infty$		4.5	5.0		5.0	7.0	mA
Transient Response								
Rise Time			50			75		ns
Overshoot			25%			25%		%
Slew Rate		1.3	1.6		1.3	1.6		V/µs
Unity Gain Bandwidth		2.8	3.5		2.8	3.5		MHz
Phase Margin	$R_L = 2\ K\Omega\ R_C = 50\ pF$		50			50		degrees
Full Power Bandwidth	$V_0 = 20V$ p-p	20	25		20	25		kHz
Input Noise Voltage	f = 20 Hz to 20 kHz		0.8	2.0		0.8	2.0	µV RMS
Input Noise Current	f = 20 Hz to 20 kHz		15			15		pA RMS
Channel Separation			108			108		dB
The following specifications apply for −55°C $\leqslant T_A \leqslant$ +125°C for RM4156, −40°C $\leqslant T_A \leqslant$ +85°C for RV4156, 0°C $\leqslant T_A \leqslant$ +70°C for RC4156.								
Input Offset Voltage	$R_S \leqslant 10\ K\Omega$			5.0			6.5	mV
Input Offset Current				75			100	nA
Input Bias Current				325			400	nA
Large Signal Voltage Gain	$R_L \geqslant 2\ K\Omega\ V_{OUT}$ ±10V	25,000			15,000			V/V
Output Voltage Swing	$R_L \geqslant 2\ K\Omega$	±10			±10			V
Supply Current			10			10		mA
Average Offset Voltage Drift			5			5		µV/°C

Notes: 1. Rating applies for case temperature of +25°C maximum; derate linearity at 6.4 mW/°C for temperatures above +25°C.
        2. For supply voltages less than ±15V, the absolute maximum input voltage is equal to the supply voltage.
        3. Short circuit to ground on one amplifier only.

Fig. 1-13 (Cont.). 4156 quad high-

TYPICAL PERFORMANCE DATA

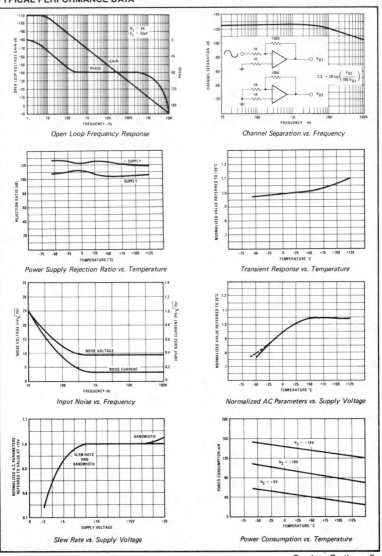

Open Loop Frequency Response

Channel Separation vs. Frequency

Power Supply Rejection Ratio vs. Temperature

Transient Response vs. Temperature

Input Noise vs. Frequency

Normalized AC Parameters vs. Supply Voltage

Slew Rate vs. Supply Voltage

Power Consumption vs. Temperature

Courtesy Raytheon Co.

performance operational amplifier.

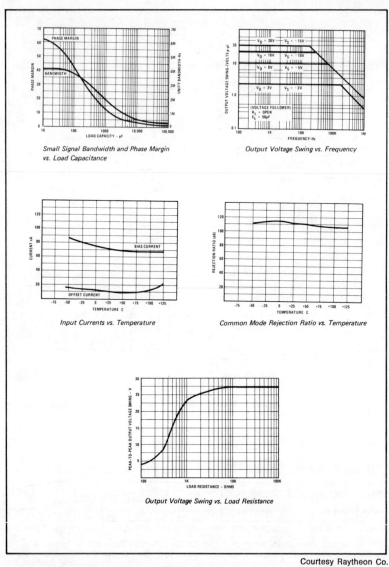

Small Signal Bandwidth and Phase Margin
vs. Load Capacitance

Output Voltage Swing vs. Frequency

Input Currents vs. Temperature

Common Mode Rejection Ratio vs. Temperature

Output Voltage Swing vs. Load Resistance

Courtesy Raytheon Co.

Fig. 1-13 (Cont.). 4156 quad high-performance operational amplifier.

EXPERIMENT NO. 1

Purpose

The purpose of this experiment is to measure the input offset voltage of a 741 op-amp.

Pin Configuration of 741 Op-Amp (Fig. 1-14)

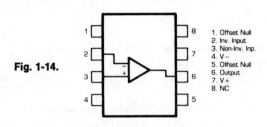

Fig. 1-14.

1. Offset Null
2. Inv. Input
3. Non-Inv. Inp.
4. V−
5. Offset Null
6. Output
7. V+
8. NC

Schematic Diagram of Circuit (Fig. 1-15)

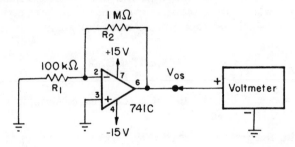

Fig. 1-15.

Design Basics

- Closed-loop gain: $A_{CL} = \dfrac{R_2}{R_1}$
- Input offset voltage: $V_{oi} = \dfrac{V_{os}}{A_{CL}}$

Step 1

Wire the circuit shown in the schematic diagram. For this experiment, the power supply connections to the op-amp are shown in the schematic diagram. For the remaining experiments they will be omitted from the diagrams, as these connections are usually implied.

Step 2

Apply power to the breadboard, and with your voltmeter (preferably a digital type), measure the output voltage, and record your result below:

$$V_{os} = \underline{\hspace{2cm}} \text{ mV}$$

I measured 56.2 mV, although the reading varied from 56.2 to 58.3 mV.

Step 3

Using the formula given in the *Design Basics* section, calculate the input offset voltage, V_{oi}, recording your result below:

$$V_{oi} = \underline{\hspace{2cm}} \text{ mV}$$

For the 741C op-amp, the typical value for the input offset voltage is 2 mV, with a maximum of 6 mV. In my case, I calculated an input offset voltage of 5.62 mV, which is still within the device's rating.

Since all the experiments in the first 7 chapters use the 741 op-amp powered by a dual voltage supply, when you are finished with each experiment, disconnect all the connections to the op-amp except the power (pins 4 and 7). In this way, you won't forget to make these connections when wiring the next experiment, as these power connections are usually omitted from schematic diagrams.

EXPERIMENT NO. 2

Purpose

The purpose of this experiment is to measure the input bias currents of the 741C op-amp.

Schematic Diagram of Circuit (Fig. 1-16)

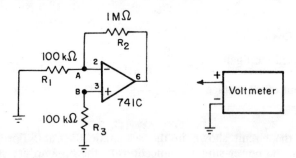

Fig. 1-16.

Design Basics

- Bias current: $I_{B1} = \dfrac{V_A}{R_1}$

$$I_{B2} = \dfrac{V_B}{R_3}$$

Step 1

Wire the circuit shown in the schematic diagram. *Don't forget the power supply connections to the op-amp!*

Step 2

Apply power to the breadboard. First measure the voltage across resistor $R_1(V_A)$, recording your result below:

$$V_A = \underline{\hspace{1cm}} mV$$

Then measure the voltage across resistor $R_3(V_B)$, recording your result below:

$$V_B = \underline{\hspace{1cm}} mV$$

For my experiment, I measured 15.2 mV for V_A and 12.7 mV for V_B.

Step 3

From the formulas given in the *Design Basics* section, calculate the input bias currents I_{B1} and I_{B2}, recording your results below:

$$I_{B1} = \underline{\hspace{1cm}} nA$$

$$I_{B2} = \underline{\hspace{1cm}} nA$$

For the *ideal* op-amp both bias currents are equal; however, for the "real world" device, the input bias currents are not equal. The manufacturer gives a value that is the average of the two. Take the average of the two currents above, and record your result below:

$$I_B \text{ average} = \underline{\hspace{1cm}} nA$$

For the 741C op-amp, the typical average value for the input bias current is 80 nA, with a maximum of 500 nA. For my experiment, I measured an average current of 139 nA, which is within the device's rating.

EXPERIMENT NO. 3

Purpose

The purpose of this experiment is to measure the intrinsic input impedance of a 741C op-amp.

Schematic Diagram of Circuit (Fig. 1-17)

Design Basics

- $Z_i = R$, when $V_i' = \frac{1}{2} V_i$

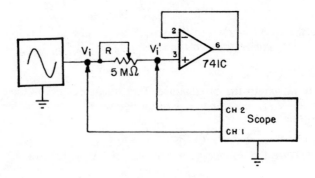

Fig. 1-17.

Step 1

Set your oscilloscope for the following settings:

- Channels 1 & 2: 0.5 volt/division
- Time base: 1 msec/division
- AC coupling

Step 2

Wire the circuit shown in the schematic diagram; again, *don't forget the power supply connections!* Apply power to the breadboard and adjust the input sine wave at 1 volt peak-to-peak, and the frequency so that 1 full cycle occupies the 10 horizontal divisions (100 Hz).

Step 3

While watching the voltage V' displayed on Channel 2, adjust the 5-MΩ potentiometer until this voltage is *one-half* the input voltage V_i (i.e., 0.5 volt peak-to-peak).

Step 4

When you have reached this point, disconnect the power to the breadboard. Without disturbing the potentiometer's setting, remove the potentiometer from the breadboard. Then measure its resistance with an ohmmeter and record its value below:

$$R_{potentiometer} = \underline{\hspace{1cm}} \Omega$$

This value is equal to the input impedance Z_i of the op-amp. For the 741C op-amp, the input impedance is typically 2 MΩ, with a minimum value of 300 kΩ. For my experiment, I measured a value of 1.2 MΩ, which is within the device's rating.

Purpose

The purpose of this experiment is to measure the slew rate of a 741C op-amp.

Schematic Diagram of Circuit (Fig. 1-18)

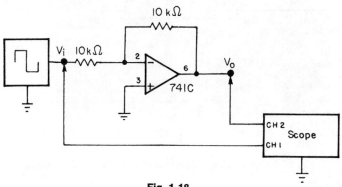

Fig. 1-18.

Step 1

Set your oscilloscope for the following settings:

- Channel 1: 5 volts/division
- Channel 2: 1 volt/division
- Time base: 10 μsec/division
- AC coupling

Step 2

Wire the circuit shown in the schematic diagram and apply power to the breadboard. Adjust the square-wave input voltage at 5 volts peak-to-peak, and the input frequency so that 1 cycle occupies the scope's display (10 kHz), as shown in Fig. 1-19.

Step 3

Measure the peak-to-peak output voltage ΔV, and record your result below:

$$\Delta V = \underline{\hspace{2cm}} \text{ volts}$$

Step 4

As illustrated in Fig. 1-19, measure the time Δt that it takes for the output voltage to switch either from its minimum to its maximum value, or vice versa, recording your result below:

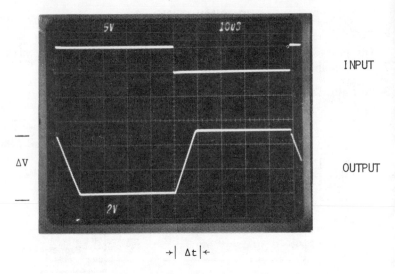

Fig. 1-19. Input and output waveforms for the 741C op-amp.

$$\Delta t = \text{_____} \mu\text{sec}$$

Step 5

From the measurements in Steps 3 and 4, calculate the slew rate $\Delta V/\Delta t$, and record your result below:

$$\text{slew rate} = \text{_____} V/\mu\text{sec}$$

For the 741C op-amp, the slew rate is typically 0.5 V/μsec. As shown in Fig. 1-19, ΔV is 4.9 volts, and Δt is 10 μsec, giving a slew rate of .49 V/μsec.

Step 6

Disconnect the power to the breadboard and insert a LM318 op-amp in place of the 741C (the pin connections are the same!). Apply power to the breadboard. What do you notice about the output waveform?

The output waveshape no longer resembles a trapezoid, but very nearly is identical to the input signal, as shown in Fig. 1-20. This is because the slew rate of the LM318 op-amp is typically 70 V/μsec, or 140 times better than the 741C! Consequently, at 10 kHz this op-amp is capable of responding to its input signal without distortion.

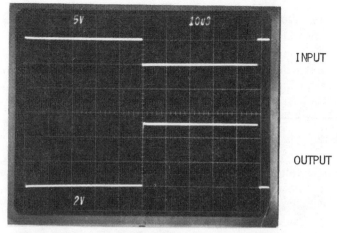

Fig. 1-20. Input and output waveforms for the LM318 op-amp.

EXPERIMENT NO. 5

Purpose

The purpose of this experiment is to determine the common-mode rejection ratio, or CMRR, of a 741C operational amplifier.

Schematic Diagram of Circuit (Fig. 1-21)

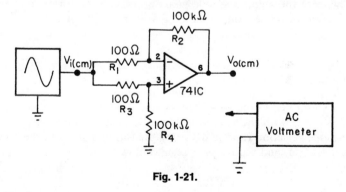

Fig. 1-21.

Design Basics

- Differential gain: $A_d = \dfrac{R_2}{R_1}$

$$= \dfrac{R_4}{R_3}$$

- Common-mode gain: $A_{CM} = \dfrac{V_{o(cm)}}{V_{i(cm)}}$

- Common-mode rejection ratio: CMRR (dB) $= 20 \log_{10} \dfrac{A_d}{A_{CM}}$

Step 1

Wire the circuit shown in the schematic diagram. Apply power to the breadboard and adjust the input frequency at approximately 60–100 Hz. If you have a small filament transformer, you can use this instead of a sine-wave generator.

Step 2

With your voltmeter measuring $V_{i(cm)}$, adjust the input voltage for at least 2 volts rms (this is an *ac* voltage measurement!). If you are using a filament transformer, this value will be approximately 6 volts rms. In either case, record the common-mode input voltage below:

$$V_{i(cm)} = \text{_____} \text{ volts rms}$$

Step 3

Now measure the corresponding common-mode output voltage, $V_{o(cm)}$, recording it below:

$$V_{o(cm)} = \text{_____} \text{ volts rms}$$

Step 4

Calculate the common-mode gain A_{CM}, using the formula given in the *Design Basics* section, and record your result below:

$$A_{CM} = \text{_____}$$

Step 5

For this circuit, called a *difference amplifier,* and whose operation will be described in the next chapter, the differential gain A_d is 1000 (i.e., R_2/R_1) with the components shown. Now determine the common-mode rejection ratio in dB using the formula given in the *Design Basics* section, and record your result below:

$$\text{CMRR} = \text{_____} \text{ dB}$$

For the 741C op-amp, the common-mode rejection ratio is typically 90 dB, with a minimum of 70 dB. To be technically correct, the term "common-mode rejection ratio" refers only to the ratio A_d/A_{CM}, while the term "common-mode rejection" refers to the common-mode rejection ratio expressed in dB. Therefore, the common-mode rejection ratio for the 741C op-amp is typically 31,600, where the common-mode rejection is 90 dB:

common-mode rejection $= 20 \log_{10}$ (common-mode rejection ratio)

$$= 20 \log_{10} (31,600)$$
$$= 90 \text{ dB}$$

Since the exact definition of these two terms varies among manufacturers, we will take the position that they mean exactly the same thing, that is, the rejection expressed in dB.

As a comparison, my results for this experiment are as follows:

$$V_{i(cm)} = 3.32 \text{ volts rms}$$
$$V_{o(cm)} = 0.285 \text{ volts rms}$$
$$A_{CM} = 0.0858$$

so that,

$$CMRR = 20 \log_{10}(A_d/A_{CM})$$
$$= 20 \log_{10}(1000/0.0858)$$
$$= 20 \log_{10}(11,655)$$
$$= 81.3 \text{ dB}$$

EXPERIMENT NO. 6

Purpose

The purpose of this experiment is to determine the closed-loop response of the 741C op-amp by calculating its gain-bandwidth product (GBP).

Schematic Diagram of Circuit (Fig. 1-22)

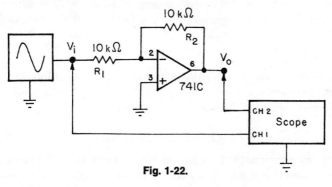

Fig. 1-22.

Design Basics

• Gain-bandwidth product: $GBP = (A_v)(BW)$

  where: $A_v = \dfrac{R_2}{R_1}$

BW = op-amp bandwidth (the high-frequency response where the output voltage decreases by a factor of 0.707)

Step 1

Set your oscilloscope for the following settings:

- Channel 1 & 2: 0.1 volt/division
- Time base: 0.5 μsec/division
- DC coupling

Step 2

Wire the circuit shown in the schematic diagram. Apply power to the breadboard and adjust the input voltage so that the *peak-to-peak output voltage* is 0.7 volt (7 vertical divisions). Make this adjustment as accurately as you can! As will be explained in Chapter 2, this op-amp circuit is an *inverting amplifier,* and, with the components shown, the voltage gain is unity (1.0). Check to see if the peak-to-peak input voltage is also 0.7 volt. It should be!

Step 3

Now slowly vary the input frequency until the peak-to-peak output voltage decreases to 0.5 volt. You may find that the output voltage increases slightly before it starts to decrease. Measure the frequency at the point where the output voltage is 0.5 volt peak-to-peak, and record it below:

$$BW = \text{_____} \text{ kHz } (A_v = 1.0)$$

This is the bandwidth of the op-amp when the voltage gain is 1.0. The high-frequency response is now a factor of 0.707 less than the low-frequency, or dc, response, and is equivalent to −3 dB.

Step 4

To compute the gain-bandwidth product GBP, multiply the bandwidth by the voltage gain, using the values of Step 3, and record your result below:

$$GBP = \text{_____} \text{ kHz}$$

For my experiment, I measured a bandwidth of 392 kHz. Consequently the GBP is also 392 kHz, since the voltage gain is 1.0.

Step 5

Now change resistor R_1 to 5 kΩ by placing another 10 kΩ resistor in *parallel* with the other 10 kΩ resistor. In addition, change Channel 2 to 0.2 volt/division, and the time base at 1 μsec/division. Vary the input frequency until the peak-to-peak output voltage is 1.0 volt (5 divisions). Measure the frequency at this point and record your result below:

$$BW = \underline{\hspace{1.5cm}} \text{ kHz } (A_v = 2.0)$$

Is this value less than the value you measured in Step 3? It should be!

Step 6

As in Step 4, compute the gain-bandwidth product for this circuit, using a voltage gain of 2.0.

$$GBP = \underline{\hspace{2cm}} \text{ kHz}$$

How does this value compare with the value that you computed in Step 4?

Within about 5%, these two values should be the same! By increasing the circuit's voltage gain, the bandwidth decreases; however, the gain-bandwidth product, which is dependent on both the voltage gain and the bandwidth, remains constant! Therefore, the gain-bandwidth product restricts the maximum high-frequency response of an op-amp circuit for a given value of voltage gain; the higher the voltage gain, the lower the bandwidth.

Step 7

Now change resistor R_1 to 1 kΩ. In addition, change Channel 2 to 1 volt/division and the time base to 5 μsec/division. Vary the input voltage until the peak-to-peak output voltage is 5 volts (5 vertical divisions). Measure the frequency at this point, and record your result below:

$$BW = \underline{\hspace{2cm}} \text{ kHz } (A_v = 10)$$

Step 8

As before, compute the gain-bandwidth product, using a voltage gain of 10.

$$GBP = \underline{\hspace{2cm}} \text{ kHz}$$

Within 5%, is there any difference between the above value and the values you determined in Steps 4 and 6?

The value should be the same in all three cases! *You should now conclude that the gain-bandwidth product is a constant value, with the inverse dependency of bandwidth and voltage gain.* Once you know the gain-bandwidth product for a particular op-amp, you immediately know the maximum voltage gain that you can have for a given input frequency, and vice versa. As with the slew rate, the GBP is a measure of the frequency response of an op-amp circuit. For example, if the GBP is 500 kHz, the maximum input frequency at which a given op-amp circuit can operate for a voltage gain of 100 is 5 kHz. However, the usual practice is to make the circuit's voltage gain about 10 to 20 times less than the value allowed by the GBP at the particular operating frequency. Suppose the gain-bandwidth product is 500 kHz and we wish to have an amplifier circuit with a voltage gain of 10. Consequently, this corresponds to a maximum voltage gain of 100 to 200. To be safe, assume that the maximum voltage gain is 100. Therefore, the maximum frequency at which we can expect this amplifier to properly function is:

$$BW = \frac{GBP}{A_v}$$
$$= \frac{500 \text{ kHz}}{100}$$
$$= 5 \text{ kHz}$$

For this particular example, the amplifier will work properly with a voltage gain of 10 at 5 kHz.

CHAPTER 2

Basic Linear Amplifier Circuits

INTRODUCTION

With this chapter, we begin the discussion of the basic op-amp circuits that form the cornerstone for *linear* applications; that is, the output signal is directly proportional to the input signal.

OBJECTIVES

At the completion of this chapter, you will be able to do the following:

- Design, compare, and predict the performance of the following op-amp circuits:
    noninverting amplifier
    inverting amplifier
    voltage follower
    summing amplifier
    difference amplifier
- Minimize the output offset voltage due to the input bias current, input offset current, and input offset voltage.
- Recognize the input and feedback elements of linear op-amp circuits.
- Determine the effects of feedback upon circuit performance.

THE NONINVERTING AMPLIFIER

As shown in Fig. 2-1, the op-amp is connected as a *noninverting* amplifier. This is because the input signal is applied to the op-amp's

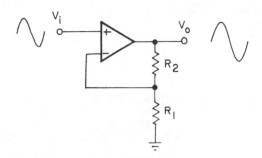

Fig. 2-1. Noninverting amplifier.

noninverting (+) input. Resistor R_1 is called the *input element,* and resistor R_2 is called the *feedback element,* since it diverts, or "feeds back" part of the output voltage to one of the op-amp's inputs. In this case, part of the output is returned to the inverting (−) input. For this noninverting amplifier, the output voltage is given by:

$$V_o = \left[1 + \frac{R_2}{R_1} \right] V_i \qquad \text{(Eq. 2-1)}$$

The *voltage gain,* or the ratio of the output voltage to the input voltage, is:

$$\text{voltage gain} = \frac{V_o}{V_i}$$

$$= 1 + \frac{R_2}{R_1} \qquad \text{(Eq. 2-2)}$$

Consequently, *the voltage gain of a noninverting amplifier will always be greater than unity (1.0)*, no matter how large we make R_1.

Since the input signal is applied to the op-amp's noninverting input, the output voltage will always be *in phase* with the input. More simply, when the input voltage goes positive, the output does the same. The only difference between the input and output voltages is that the output voltage will be $1 + R_2/R_1$ times larger than the input.

As pointed out in Chapter 1, the open-loop gain A_{OL} is an intrinsic characteristic of the op-amp when there is no feedback. When feedback is used, we then refer to the *closed-loop gain* A_{CL}, which is simply the voltage gain of the op-amp configuration (Equation 2-2), or

$$A_{CL} = 1 + \frac{R_2}{R_1} \qquad \text{(Eq. 2-3)}$$

for the noninverting amplifier. The *loop gain,* A_L is the reduction of the open-loop gain by the closed-loop gain, so that by definition,

$$A_L = \frac{A_{OL}}{A_{CL}}$$

$$= \frac{A_{OL}}{1 + \frac{R_2}{R_1}} \qquad \text{(Eq. 2-4)}$$

For all practical purposes, the input impedance of the noninverting amplifier is the *intrinsic input impedance* of the op-amp itself, which is high enough to minimize loading of the input circuitry. On the other hand, the output impedance of the circuit of Fig. 2-1 is determined from the formula:

$$Z_o = \frac{Z_{oi}}{A_L}$$

$$= Z_{oi} \left[\frac{1 + \frac{R_2}{R_1}}{A_o} \right] \qquad \text{(Eq. 2-5)}$$

where Z_{oi} is the intrinsic output impedance of the op-amp, as determined from the manufacturer's data sheet.

Example

To see how these concepts just discussed fit into place, assume that a type 741 op-amp is used in the circuit of Fig. 2-1 with $R_1 = 1$ kΩ and $R_2 = 100$ kΩ

From Equation 2-3, the voltage, or closed-loop gain is:

$$A_{CL} = 1 + \frac{R_2}{R_1}$$

$$= 1 + \frac{100 \text{ k}\Omega}{1 \text{ k}\Omega}$$

$$= 101$$

From a given manufacturer's data sheet, the open-loop gain for the 741 op-amp is typically 200,000. From Equation 2-4, the circuit's loop gain is then:

$$A_L = \frac{A_{OL}}{A_{CL}}$$

$$= \frac{200,000}{101}$$

$$\simeq 2000$$

Using a typical value of 75 Ω for the 741's intrinsic output impedance (Z_{oi}), the output impedance of the completed noninverting amplifier is then:

$$Z_o = \frac{Z_{oi}}{A_L}$$

$$= \frac{75\ \Omega}{2000}$$

$$= 0.04\ \Omega$$

From this example it should now be evident that the result of adding feedback increases the loop gain, which in turn decreases the output impedance of the amplifier circuit! We can then connect almost any load to the output, as long as the maximum output current rating of the op-amp is not exceeded.

Usually, the calculation of the op-amp circuit's output impedance can be ignored since it is such a small number. It was included here to demonstrate the effect of external feedback.

THE INVERTING AMPLIFIER

As shown in Fig. 2-2, the op-amp is connected as an *inverting* amplifier. This is because the input signal is applied to the op-amp's inverting ($-$) input through R_1, which is called the *input element*. Resistor R_2 is the *feedback element*. For the inverting amplifier, the output voltage is given by the equation

$$V_o = -\left[\frac{R_2}{R_1}\right]V_i \qquad \text{(Eq. 2-6)}$$

The minus ($-$) sign in the above equation indicates that when the input signal voltage goes positive, the output voltage goes negative, and vice versa. In other words, the output signal is of opposite polarity with respect to the input, which is the same as saying that the output is $180°$ out of phase with the input.

The voltage, or closed-loop gain, is then:

$$A_{CL} = \frac{V_o}{V_i}$$

$$= -\frac{R_2}{R_1} \qquad \text{(Eq. 2-7)}$$

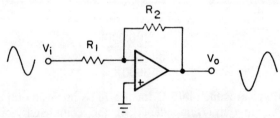

Fig. 2-2. Inverting amplifier.

Consequently, the voltage gain of the inverting amplifier can be either *less than, equal to,* or *greater than 1,* depending on the relation of R_2 to R_1.

The loop gain is then:

$$A_L = \frac{A_{OL}}{A_{CL}}$$

$$= A_{OL}\left[\frac{R_2}{R_1}\right] \qquad \text{(Eq. 2-8)}$$

Unlike the noninverting amplifier, the input impedance of the inverting amplifier circuit is simply the value of the input element, R_1, and will be much less than that of the noninverting circuit. The circuit's output impedance, as before, is determined solely by the op-amp's intrinsic output impedance and the circuit's loop gain, so that:

$$Z_o = \frac{Z_{oi}}{A_L}$$

$$= Z_{oi}\left[\frac{R_1}{A_{OL}R_2}\right] \qquad \text{(Eq. 2-9)}$$

For the special case when R_1 and R_2 are equal, we then have a *unity gain inverter,* which is handy when we only want to invert the polarity of the input signal.

The closed-loop gain of the basic inverting amplifier can be controlled by switching in different feedback resistors, as shown in Fig. 2-3. On the other hand, by using an analog switch, such as the 4016 (CMOS), we are then able to digitally control the op-amp's voltage gain as shown in Fig. 2-4. The 4016 CMOS switch typically has an

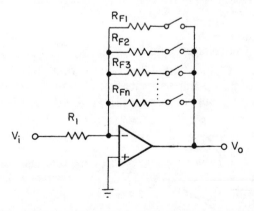

Fig. 2-3. Using different feedback resistors to control closed-loop gain of inverting amplifier.

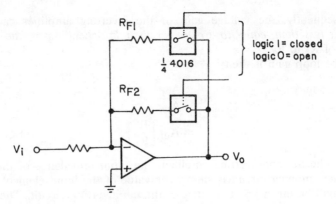

Fig. 2-4. Using an analog switch to digitally control the voltage gain of an op-amp.

ON resistance of $300\,\Omega$, which should be added to the resistance of the feedback resistance when calculating the voltage gain for each feedback loop. If the ON resistance of the 4016 poses some problems, a 4066 CMOS switch having a typical ON resistance of $80\,\Omega$ can be used instead. The analog switch can then be controlled by shift registers, counters, decoders, etc.

DC OUTPUT OFFSET

In the ideal op-amp, the output voltage is zero when the input voltage is also zero. However, all commercial op-amps have a small, but finite, dc output voltage called the output *offset voltage,* even though the input may be grounded. The dc output offset voltage is a result of three sources:

- input offset current
- input bias current
- input offset voltage

As stated in Chapter 1, the input bias current must be supplied to both inputs of the op-amp to assure that the op-amp behaves properly. For the inverting amplifier circuit of Fig. 2-5, the input bias current,

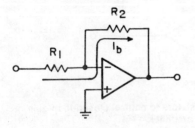

Fig. 2-5. Input bias currents for the inverting amplifier.

with no input signal, flows through both the input and feedback resistors. By Ohm's law, when a current flows through a known resistance, a voltage is developed across this resistance ($V = IR$). Since the noninverting input of the op-amp is grounded, the voltages developed across these resistors appear as a dc input voltage, which in turn is amplified by the op-amp. For the inverting amplifier circuit of Fig. 2-5, the output voltage (V_{os}) generated as a result of the input bias current (I_b) is:

$$V_{os} = I_b R_2 \qquad \text{(Eq. 2-10)}$$

The method commonly used to correct for the output voltage offset due to the input bias current is to place an additional resistor R_3 between the noninverting ($+$) input and ground as shown in Fig. 2-6.

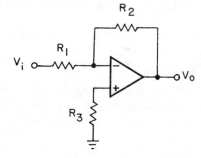

Fig. 2-6. Correcting for the output voltage offset due to input bias current.

The value of this additional resistor is equal to the parallel combination of R_1 and R_2, or:

$$R_3 = \frac{R_1 R_2}{R_1 + R_2} \qquad \text{(Eq. 2-11)}$$

so that the voltage developed across R_3 is equal and opposite to the voltage across the parallel combination of R_1 and R_2. Since the two voltages are equal and opposite, they cancel.

However, the above discussion assumes that the bias currents flowing into both inputs are equal. Unfortunately, in a typical op-amp, *both bias currents are not exactly equal,* the value for I_b (from the data sheet) being only an *average of the two input bias currents.*

Since there will be a difference in the two bias currents, called the *input offset current,* I_{os}, there will still exist a small but finite dc output offset voltage, equal to:

$$V_{os} = I_{os} R_2 \qquad \text{(Eq. 2-12)}$$

The remaining source of output offset is due to the op-amp's *input offset voltage,* resulting from mismatches in the internal circuitry and fabrication of the op-amp. As shown for the inverting amplifier circuit

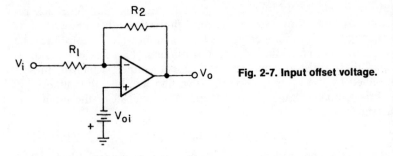

Fig. 2-7. Input offset voltage.

of Fig. 2-7, the input offset voltage , V_{oi}, can be represented as a small battery in series with the noninverting input of an ideal op-amp.

For the circuit in Fig. 2-7, the dc output offset voltage, as a result of the input offset voltage, is calculated from:

$$V_{os} = \left[1 + \frac{R_2}{R_1} \right] V_{oi} \qquad \text{(Eq. 2-13)}$$

Example

Using the inverting amplifier shown in Fig. 2-8 with R_1 grounded (i.e., zero input voltage), the following typical parameters are given for the type 741 op-amp (Signetics Analog Manual):

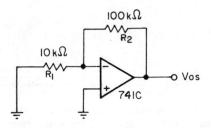

Fig. 2-8. Calculation of output offset voltage.

$$V_{oi} = 2 \text{ mV}$$
$$I_{os} = 20 \text{ nA}$$
$$I_b = 80 \text{ nA}$$

The dc output offset voltage as a result of the input bias current from Equation 2-10, is:

$$\begin{aligned} V_{os} &= I_b R_2 \\ &= (80 \times 10^{-9} \text{ A})(100 \times 10^3 \text{ }\Omega) \\ &= 8 \text{ mV} \end{aligned}$$

The dc output offset voltage as a result of the input offset current from Equation 2-12, is then:

$$V_{os} = I_{os}R_2$$
$$= (20 \times 10^{-9}\,\text{A})(100 \times 10^3\,\Omega)$$
$$= 2\,\text{mV}$$

Finally, the dc output offset voltage as a result of the input offset voltage, from Equation 2-12, is then:

$$V_{os} = \left[1 + \frac{R_2}{R_1}\right]V_{oi}$$
$$= (11)(2\,\text{mV})$$
$$22\,\text{mV}$$

Without the addition of R_3, as shown in Fig. 2-6, the dc output offset is simply the contribution of the bias current offset and the input voltage offset, which can range from 14 mV to 30 mV, depending on whether the two components add or subtract.

If we add a 9.1-kΩ resistor of R_3 in series with the noninverting input as shown in Fig. 2-9, the dc offset voltage is then the contribution of the input offset current and the input offset voltage. The output offset can range from 20 mV to 24 mV.

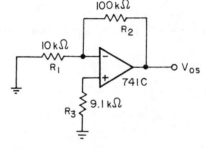

Fig. 2-9. 9.1-kΩ resistor in series with noninverting input.

Until now, only a method that corrects for the dc output offset voltage due to *input bias currents* has been discussed. From the previous example, it seems that we should be concerned with minimizing the offset due to the input offset voltage. To cancel the dc offset caused by the input offset voltage, the inverting amplifier circuit of Fig. 2-10 can be used.*

For noninverting amplifiers, a similar circuit, shown in Fig. 2-11, can be used.

However, many op-amps, such as the 741, 747, 748, and others, have provisions for cancelling the dc output offset with a single potentiometer, as shown in Fig. 2-12. The potentiometer is then used to null the output voltage when the input is grounded, or zero.

---

* Signetics Analog Manual, 1976, p. 61.

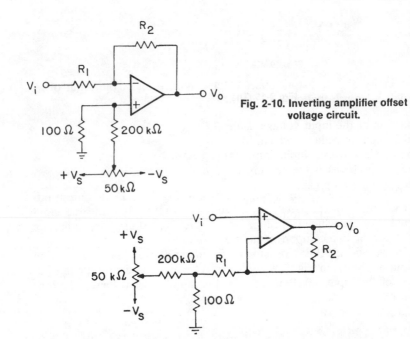

Fig. 2-10. Inverting amplifier offset voltage circuit.

Fig. 2-11. Noninverting amplifier offset voltage circuit.

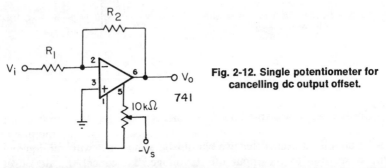

Fig. 2-12. Single potentiometer for cancelling dc output offset.

As a general rule, offset cancellation circuits are not required for the circuits discussed in this chapter, but are included here as "methods of last resort." When such methods are required, they will be pointed out in this book.

THE VOLTAGE FOLLOWER

The *voltage follower,* or *source follower,* shown in Fig. 2-13, is simply a unity-gain noninverting amplifier. There is no input or feedback resistance. For this circuit, the output voltage is an exact repro-

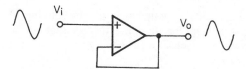

Fig. 2-13. The voltage follower.

duction of the input voltage. Like the noninverting amplifier, it has a high input impedance which, for all practical purposes, is equal to the op-amp's intrinsic input impedance. The output impedance for the voltage follower is the op-amp's output impedance divided by the op-amp's open-loop gain.

The function of the voltage follower is identical to the cathode, emitter, and source followers for vacuum tubes, bipolar transistors, and field-effect transistors, respectively. That is, the circuit is used to buffer an input signal from its load, since its input impedance is high and its output impedance is low.

THE SUMMING AMPLIFIER

If several input resistors are simultaneously connected to the op-amp's inverting input, as shown in Fig. 2-14, we have a *summing amplifier*. For this circuit, the separate input voltages are added, so that the output voltage is:

$$V_o = -\left[\frac{R_F}{R_1}V_1 + \frac{R_F}{R_2}V_2 + \frac{R_F}{R_3}V_3\right]$$
$$= -R_F\left[\frac{V_1}{R_1} + \frac{V_2}{R_2} + \frac{V_3}{R_3}\right] \qquad \text{(Eq. 2-14)}$$

which is similar to Equation 2-6, except that we now have multiple inputs. If the feedback resistance R_F and the input resistances R_1, R_2, and R_3 are made equal, then the output voltage will be:

$$V_o = -(V_1 + V_2 + V_3) \qquad \text{(Eq. 2-15)}$$

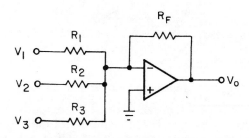

Fig. 2-14. The summing amplifier.

On the other hand, we can have gain by making the value of the feedback resistance larger than the three equal input resistors. The input impedance for each input is simply the value of the corresponding input resistor.

One nice variation of this circuit is the *averager*. By suitable selection of the ratio of the feedback resistance to the equal input resistances, we are then able to take the average value of the input voltages. Taking the circuit of Fig. 2-14 for example, the average of the three inputs must be:

$$(V_o)_{average} = \frac{V_1 + V_2 + V_3}{3}$$

If we make $R_1 = R_2 = R_3 = R_F/3$, we then are able to obtain the average of these three input signals.

Example

Design a circuit that will take the average of two input voltages.

As shown in Fig. 2-15, we have the same circuit as Fig. 2-14 except that there are only two inputs. In order to obtain the average of these two inputs, we must then make the feedback resistor R_F *one-half* the value of the two equal input resistors R_1 and R_2. If R_1 and R_2 are both 20 kΩ, then R_F must be 10 kΩ. Using Equation 2-14,

$$V_o = -R_F \left[\frac{V_1}{R_1} + \frac{V_2}{R_2} \right]$$

$$= -(10 \text{ k}\Omega) \left[\frac{V_1}{20 \text{ k}\Omega} + \frac{V_2}{20 \text{ k}\Omega} \right]$$

$$= -\left[\frac{V_1 + V_2}{2} \right]$$

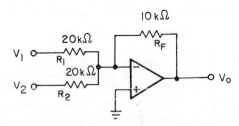

Fig. 2-15. A summing amplifier with two inputs.

Again, the minus $(-)$ sign tells us that the output signal will be of opposite polarity with respect to the average of the two input signals.

For critical applications, the dc output offset voltage due to the input bias currents can be minimized for the summing amplifier cir-

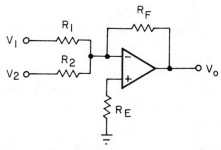

Fig. 2-16. Resistor connected in series with noninverting input of summing amplifier.

cuit of Fig. 2-15. As with the simple inverting amplifier circuit of Fig. 2-6, a resistor, equal to the parallel combination of the input and feedback resistors, is placed between the op-amp's noninverting input and ground, as shown in Fig. 2-16. For this circuit, the value of this resistor R_E must be:

$$R_E = \frac{1}{\dfrac{1}{R_1} + \dfrac{1}{R_2} + \dfrac{1}{R_F}} \qquad \text{(Eq. 2-16)}$$

Using the component values of the previous example, the value for R_E is:

$$R_E = \frac{1}{\dfrac{1}{20 \text{ k}\Omega} + \dfrac{1}{20 \text{ k}\Omega} + \dfrac{1}{10 \text{ k}\Omega}}$$

$$= \frac{1}{20 \times 10^{-4}}$$

$$= 5 \text{ k}\Omega$$

However, this resistor can be omitted for most applications.

THE DIFFERENCE AMPLIFIER

As shown in Fig. 2-17, the difference amplifier has input voltages that are applied simultaneously to the inverting and noninverting inputs of the op-amp. Although this type of circuit looks complex, the analysis is quite simple using what we have already encountered in this chapter. First, assume that the point V_2' is shorted to ground and V_2 is zero. We now have a configuration that is identical to Fig. 2-2, or a simple inverting amplifier. The output voltage is then:

$$V_o = -\left[\frac{R_F}{R_1}\right]V_1 \qquad \text{(Eq. 2-17)}$$

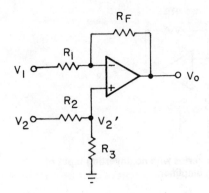

Fig. 2-17. The difference amplifier.

Next, remove the short circuit at V_2' and then short the input signal V_1 to ground. We essentially have a noninverting amplifier. The actual noninverting input voltage that the op-amp sees is V_2' which is related to the input voltage V_2 by the voltage-divider relation:

$$V_2' = \left[\frac{R_3}{R_2 + R_3}\right]V_2 \qquad \text{(Eq. 2-18)}$$

The noninverting output voltage is then:

$$V_o = \left[1 + \frac{R_F}{R_1}\right]V_2'$$
$$= \left[1 + \frac{R_F}{R_1}\right]\left[\frac{R_3}{R_2 + R_3}\right]V_2 \qquad \text{(Eq. 2-19)}$$

Combining Equations 2-17 and 2-19, the output voltage for the difference amplifier as a function of the input voltages V_1 and V_2 is:

$$V_o = -\left[\frac{R_F}{R_1}\right]V_1 + \left[1 + \frac{R_F}{R_1}\right]\left[\frac{R_3}{R_2 + R_3}\right]V_2 \quad \text{(Eq. 2-20)}$$

The first right-hand term is the *inverted* output, while the second term is the *noninverted* output.

When the circuit of Fig. 2-17 is used as a difference amplifier, the voltage gain (G) is set by all four resistors, so that:

$$R_F = GR_1 \qquad \text{(Eq. 2-21)}$$
$$R_3 = GR_1 \qquad \text{(Eq. 2-22)}$$
$$R_2 = R_1 \qquad \text{(Eq. 2-23)}$$

When all four resistors are equal, Equation 2-20 reduces to:

$$V_o = V_2 - V_1 \qquad \text{(Eq. 2-24)}$$

so that the output voltage is the difference of V_2 and V_1. Such a circuit is then called a unity-gain *analog subtractor*.

Example

Design a difference amplifier with a voltage gain of 10, using the circuit of Fig. 2-17.

From Equations 2-21, 2-22, and 2-23 letting $R_1 = 10$ kΩ, we find that:

$$R_F = 100 \text{ k}\Omega$$
$$R_3 = 100 \text{ k}\Omega$$
$$R_2 = 10 \text{ k}\Omega$$

so that the completed circuit is shown below in Fig. 2-18.

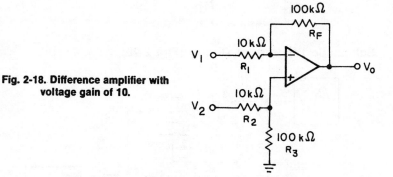

Fig. 2-18. Difference amplifier with voltage gain of 10.

AN INTRODUCTION TO THE EXPERIMENTS

The following experiments are designed to demonstrate the design and operation of the fundamental linear amplifier circuits whose output signal is directly proportional to the input.

The experiments that you will perform can be summarized as follows:

Experiment No.	Purpose
1	Demonstrates the design and operation of a voltage follower.
2	Demonstrates the design and operation of a noninverting amplifier.
3	Demonstrates the design and operation of an inverting amplifier.
4	Demonstrates the design and operation of a 2-input summing amplifier.
5	Demonstrates the design and operation of a difference amplifier.

EXPERIMENT NO. 1

Purpose

The purpose of this experiment is to demonstrate the operation of voltage follower, using a type 741 op-amp.

Pin Configuration of 741 Op-Amp (Fig. 2-19)

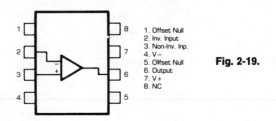

1. Offset Null
2. Inv. Input
3. Non-Inv. Inp.
4. V−
5. Offset Null
6. Output
7. V+
8. NC

Fig. 2-19.

Schematic Diagram of Circuit (Fig. 2-20)

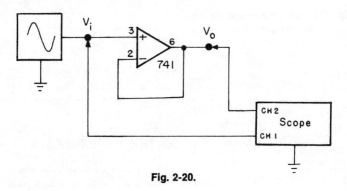

Fig. 2-20.

Design Basics

- Voltage gain: $\dfrac{V_o}{V_i} = 1$

Step 1

Set your oscilloscope for the following settings:

- Channels 1 & 2: 0.5 volt/division
- Time base: 1 msec/division
- AC coupling

Step 2

First check your wired circuit, making sure that it is correct. Don't forget the +V and −V power supply connections, as they are usually omitted from schematic diagrams! Pin 7 goes to +V and pin 4 goes to −V. Apply power to the breadboard and observe the input and output traces on the screen of the scope.

NOTE: Since we will be concerned with both the input and output signals, we will adopt the convention that the input signal is Channel 1, and the output signal is Channel 2. When viewing both signals simultaneously on a dual-trace oscilloscope, *position the input signal so that it is above the output signal.*

Step 3

Adjust the output of the generator so that the voltage is 1.5 volts peak-to-peak (3 vertical divisions), and the generator frequency so that there are at least 4 complete cycles on the oscilloscopes screen (at least 400 Hz). What is the difference between the input and output signals?

There is *no difference* between the two signals, as they are in phase. The output voltage is also 1.5 volts peak-to-peak. Consequently, the voltage gain of this voltage follower is 1.0, which is always the case.

Step 4

Verify that the voltage gain of a voltage follower is always equal to 1 by randomly varying the input voltage and measuring the corresponding output voltage.

EXPERIMENT NO. 2

Purpose

The purpose of this experiment is to demonstrate the operation of a noninverting amplifier, using a type 741 op-amp.

Schematic Diagram of Circuit (Fig. 2-21)

Design Basics

- Voltage gain $= \dfrac{V_o}{V_i}$

$$= 1 + \dfrac{R_B}{R_A}$$

Step 1

Set the oscilloscope for the following settings:

- Channels 1 & 2: 0.5 volt/division
- Time base: 1 msec/division
- AC coupling

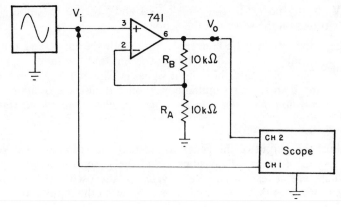

Fig. 2-21.

Step 2

Apply power to the breadboard and adjust the generator's output voltage at 1 volt peak-to-peak and the frequency at 400 Hz (4 complete cycles). With the amplifier's input signal positioned *above* the output signal on the oscilloscope's screen, what is the difference between the two signals?

The only difference between the two signals is that the output signal is *larger* than the input signal, as shown in Fig. 2-22. Both signals are

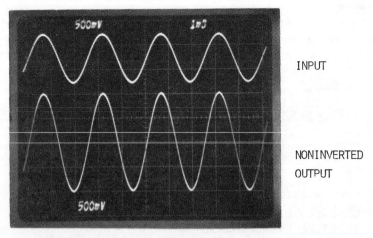

INPUT

NONINVERTED
OUTPUT

Fig. 2-22. Input and output waveforms for a noninverting amplifier.

said to be *in phase,* since the output signal goes positive exactly when the input does.

Step 3

What is the peak-to-peak output voltage?

You should have measured approximately 2 volts peak-to-peak. What then is the voltage gain?

The voltage gain is 2.0. How does this compare with the equation given in the *Design Basics* section?

By the equation

$$\text{Voltage gain} = 1 + \frac{R_B}{R_A}$$
$$= 1 + \frac{10 \text{ k}\Omega}{10 \text{ k}\Omega}$$
$$= 2.0$$

Step 4

Keeping the input level constant at 1 volt peak-to-peak, change resistor R_B, and complete the following table. Do your experimental results agree with the design equation?

R_B	Measured V_o (peak-to-peak)	Voltage Gain
27 kΩ		
39 kΩ		
47 kΩ		
82 kΩ		

EXPERIMENT NO. 3

Purpose

The purpose of this experiment is to demonstrate the operation of the inverting amplifier, using the type 741 op-amp.

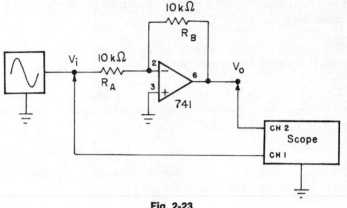

Fig. 2-23.

Schematic Diagram of Circuit (Fig. 2-23)

Design Basics

- Voltage gain $= \dfrac{V_o}{V_i}$

$$= -\dfrac{R_B}{R_A}$$

Step 1

Set the oscilloscope for the following settings:

- Channels 1 & 2: 0.5 volt/division
- Time base: 1 msec/division
- AC coupling

Step 2

Apply power to the breadboard and adjust the generator's output voltage at 1 volt peak-to-peak and the frequency so that there are about 5 complete cycles for the 10 horizontal divisions (500 Hz). With the circuit's input voltage positioned above the output voltage on the oscilloscope's screen, what is the difference between the two signals?

The output signal is of oppositive form, or is *inverted,* compared with the input signal, as shown in Fig. 2-24. The output is said to be inverted, or 180° out-of-phase with the input, since the positive peak of the output signal occurs when the input's peak is negative.

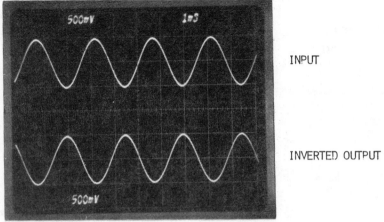

500mV   1mS

INPUT

INVERTED OUTPUT

500mV

Fig. 2-24. Input and output waveforms for an inverting amplifier.

Step 3

What is the peak-to-peak output voltage?

The peak-to-peak output voltage should be 1 volt, which is the same as the input. Consequently, the voltage gain is -1.0, where the minus sign indicates that the output is inverted with respect to the input. Also, by the equation:

$$\text{voltage gain} = -\frac{R_B}{R_A}$$
$$= -\frac{10 \text{ k}\Omega}{10 \text{ k}\Omega}$$
$$= -1.0$$

Step 4

Keeping the input level constant at 1 volt peak-to-peak, change resistor R_B, and complete the following table. Do your experimental results agree with the design equation?

R_B	Measured V_o (peak-to-peak)	Voltage Gain
27 kΩ		
39 kΩ		
47 kΩ		
82 kΩ		

Purpose

The purpose of this experiment is to demonstrate the operation of a 2-input summing amplifier, using a type 741 op-amp.

Schematic Diagram of Circuit (Fig. 2-25)

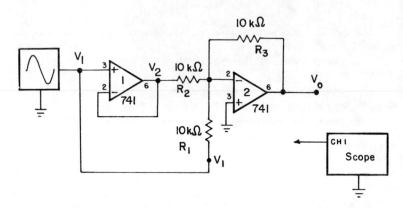

Fig. 2-25.

Design Basics

- Output voltage: $V_o = -R_3 \left[\dfrac{V_1}{R_1} + \dfrac{V_2}{R_2} \right]$

Step 1

Set the oscilloscope for the following settings:

- Channel 1: 1 volt/division
- Time base: 1 msec/division
- AC coupling

Step 2

Apply power to the breadboard and adjust the peak-to-peak output voltage of the function generator (V_1) at 1 volt, and adjust the frequency so that there are about 3 full cycles on the scope's screen (300 Hz).

Step 3

Measure the output voltage at the output of the 1st op-amp (V_2). What is it?

You should have measured a peak-to-peak voltage of 1 volt, since this portion of the circuit is just a voltage follower whose operation was described in Experiment No. 1.

Step 4

Measure the voltage at the output of the 2nd op-amp (V_o). What is it?

You should have measured a peak-to-peak voltage of approximately 2.0 volts. Why?

This 2nd amplifier is the *summing amplifier,* adding the two input voltages V_1 (1 volt) and V_2 (also 1 volt). This can be verified by the equation in the *Design Basics* section so that:

$$V_o = -R_3\left[\frac{V_1}{R_1}+\frac{V_2}{R_2}\right]$$
$$= -\left[\frac{10\text{ k}\Omega}{10\text{ k}\Omega}(1\text{ volt}) + \frac{10\text{ k}\Omega}{10\text{ k}\Omega}(1\text{ volt})\right]$$
$$= -2.0\text{ volts}$$

The negative sign occurs because we are using the op-amp as an inverting amplifier, so that the output is inverted with respect to the sum of the two inputs which are in phase.

If we are able to simultaneously observe V_1, V_2, and V_o on the oscilloscope's screen, the three traces would look like Fig. 2-26.

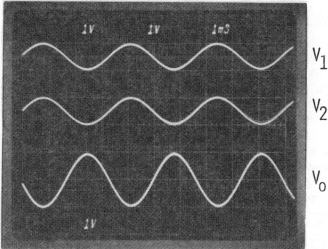

Fig. 2-26. Waveforms for the summing amplifier in Fig. 2-25.

Step 5

So far we have only presented the simple case of adding two equal voltages. To demonstrate that the equation in Step 4 and the operation of the summing amplifier still hold for *unequal* input voltages, disconnect the power from the breadboard and *rewire only the 1st op-amp as a noninverting amplifier,* as shown in Fig. 2-27. The 2nd op-amp remains connected as before.

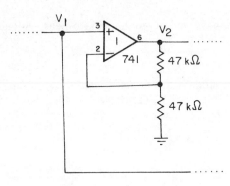

Fig. 2-27.

Step 6

Apply power again to the breadboard. What is V_2 now (i.e., the output voltage of the new circuit for the 1st op-amp)? Is it what you expected?

You should have measured approximately 2 volts, since the voltage gain of this noninverting amplifier is 2.0.

Step 7

Now measure V_o (the output voltage of the 2nd op-amp). What is it?

The peak-to-peak voltage should be approximately 3 volts. If we were again able to simultaneously observe V_1, V_2, and V_o, the three traces would look like Fig. 2-28.

Step 8

Again, disconnect the power supply and rewire the 1st op-amp as a unity-gain inverting amplifier, as shown in Fig. 2-29.

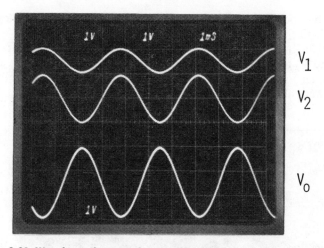

Fig. 2-28. Waveforms for summing amplifier with 1st op-amp connected as a noninverting amplifier.

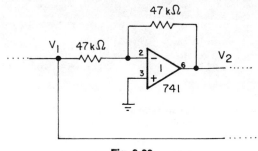

Fig. 2-29.

Step 9

Apply power to the breadboard and now measure V_o. What do you get?

You should measure no output voltage! Why?

You would probably think that the output voltage (V_o) would be 2 volts, since V_1 and V_2 are now each 1 volt. I have played a little trick on you. In Step 8 we were using a *unity-gain amplifier,* so that the output voltage (V_2) was inverted with respect to its input, V_1. When these two equal, but out-of-phase voltages were added, they cancelled each other, resulting in a net output of zero. This can be

75

seen by looking at V_1, V_2, and V_o simultaneously, as shown in Fig. 2-30.

When V_1 goes positive, V_2 goes negative by an equal amount. When V_1 and V_2 are added, the net result is zero. The same analysis applies for when V_1 goes negative. In Steps 1 through 7, the two input voltages were always in phase.

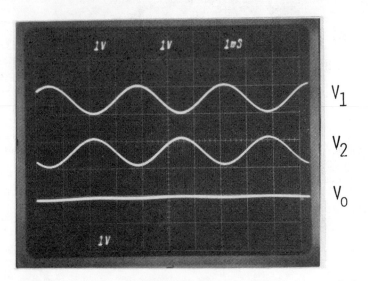

Fig. 2-30. Waveforms for summing amplifier with 1st op-amp connected as a unity-gain inverting amplifier.

EXPERIMENT NO. 5

Purpose

The purpose of this experiment is to demonstrate the design and operation of an op-amp difference amplifier, using a type 741 op-amp.

Schematic Diagram of Circuit (Fig. 2-31)

Design Basics

- $V_o = \dfrac{R_2}{R_1}(V_B - V_A)$

  when: $R_1 = R_3$

  $ R_2 = R_4$

Step 1

Wire the circuit shown in the schematic diagram and then apply power to the breadboard.

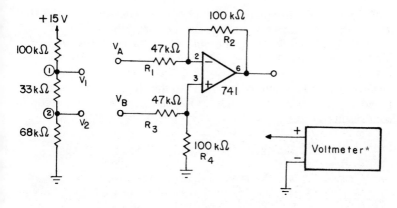

Fig. 2-31.

Step 2

First connect the noninverting input resistor (R_3) to point 1 and the inverting resistor (R_1) to point 2 on the resistor divider string.

Step 3

With your voltmeter, measure the dc input voltages V_1 (V_B) and V_2 (V_A), recording your results below:

$$V_1 = V_B = \text{_____} \text{ volts}$$

$$V_2 = V_A = \text{_____} \text{ volts}$$

Then take the difference, $V_B - V_A$, recording your result below:

$$V_B - V_A = \text{_____} \text{ volts}$$

Step 4

Now with your voltmeter, measure the output voltage V_o and record your result below:

$$V_o = \text{_____} \text{ volts}$$

Step 5

From the equation given in the Design Basics section, compare the output voltage as calculated from this equation with the output voltage you measured in Step 4. They should be nearly the same.

---

* Due to the loading error that would be introduced by using a volt-ohm-milliammeter (vom), it is suggested that a vtvm or dvm with high input impedance (10 megohms or more) be used.

From the design equation, the *difference gain* A_d is 2.128 (i.e., R_2/R_1), so that the output voltage is 2.128 times the quantity $V_B - V_A$. Since V_B is greater than V_A, the polarity of the output is *positive*.

Step 6

Now reverse the input connections so that R_1 is connected to point 1 and R_3 is connected to point 2. Repeat Steps 3 and 4, recording your results below:

$$V_1 = V_A = \underline{\hspace{1.5cm}} \text{ volts}$$

$$V_2 = V_B = \underline{\hspace{1.5cm}} \text{ volts}$$

$$V_B - V_A = \underline{\hspace{1.5cm}} \text{ volts}$$

$$V_o = \underline{\hspace{1.5cm}} \text{ volts}$$

Since V_B is now less than V_A, the output voltage should have a negative polarity. Otherwise, the magnitude of the output voltage for both connections should approximately be the same. Therefore, the following rules for the operation of this difference amplifier can be stated as:

1. If V_B is greater than V_A, the polarity of the output voltage will be positive.
2. If V_B is less than V_A, the polarity of the output voltage will be negative.

Step 7

Disconnect the power from the breadboard and change resistors R_1 and R_3 to 33 kΩ. The difference gain is now equal to 3.3. Repeat this experiment for the new gain.

CHAPTER 3

The Differentiator and Integrator

INTRODUCTION

In this chapter, two additional basic op-amp circuits are presented: the differentiator and integrator, which are mathematical inverses of each other.

OBJECTIVES

At the completion of this chapter, you will be able to do the following:

- Design circuits for and predict the performance of an:
  - op-amp differentiator
  - op-amp integrator
- Recognize the limitations of these circuits.

THE DIFFERENTIATOR

As shown in the circuit of Fig. 3-1, the basic op-amp differentiator (not to be confused with the difference amplifier) is similar to the basic inverting amplifier, except that the input element is a capacitor. For this circuit, the output voltage is given by:

$$V_o = -R_F C \frac{\Delta V_i}{\Delta t} \qquad \text{(Eq. 3-1)}$$

where the quantity $\Delta V_i / \Delta t$ is the change in input voltage over a specified time interval. Such a quantity is often referred to as the *slope of the line*. Using calculus, Equation 3-1 can be rewritten as:

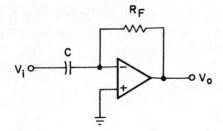

Fig. 3-1. The basic op-amp differentiator.

$$V_o = -R_F C \frac{dV_i}{dt} \qquad \text{(Eq. 3-2)}$$

One problem with the basic circuit is that the capacitor's reactance (i.e., $1/2\pi f C$) varies inversely with frequency. Consequently, the output voltage of the differentiator increases with frequency, making the circuit susceptible to high-frequency noise. A more practical differentiator circuit is shown in Fig. 3-2 with a resistor placed in series with

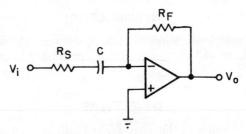

Fig. 3-2. Differentiator with limited high-frequency gain.

the input capacitor to decrease the high-frequency gain to the ratio of R_F/R_S. The output voltage as a function of time is still given by Equation 3-2. However, the circuit functions as a differentiator only for input frequencies less than:

$$f_c = \frac{1}{2\pi R_S C} \qquad \text{(Eq. 3-3)}$$

For input frequencies greater than that given by Equation 3-3, the circuit approaches an inverting amplifier with a voltage gain of

$$\frac{V_o}{V_i} = -\frac{R_F}{R_S} \qquad \text{(Eq. 3-4)}$$

In Equation 3-2, the product $R_F C$, called the *time constant,* should approximately be equal to the period of the input signal to be differentiated. In practice, R_S is usually 50–100 Ω.

Example

Design a circuit that will differentiate a 500-Hz input signal, with the circuit's high-frequency gain limited to 10.

Since the period of the input signal is 1/500 Hz, or 2 msec, then:

$$0.002 \text{ sec} = R_F C$$

Choosing $C = 1 \ \mu F$, then R_F must be 2 kΩ. Since the high-frequency gain is limited to 10, then from Equation 3-4 R_S must be 200 Ω, giving the final circuit shown in Fig. 3-3.

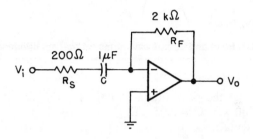

Fig. 3-3. 500-Hz differentiator with high-frequency gain limited to 10.

If the input signal is known, what will the output signal look like?

For the sine wave,

$$V_i = V_m \sin(\omega t) \qquad \text{(Eq. 3-5)}$$

where,

V_m = peak voltage of sine wave,
ω = input frequency, radian/sec, equal to $2\pi f$.

The output voltage as a function of time is then:

$$V_o = - R_F C \frac{d}{dt} (V_m \sin(\omega t))$$

$$= - \omega R_F C V_m \cos(\omega t) \qquad \text{(Eq. 3-6)}$$

The output is then a *cosine-wave,* which is nothing more than a sine wave that is 90° out-of-phase, or shifted by ¼ cycle, as shown in Fig. 3-4. From Equation 3-6 and Fig. 3-4, the peak output voltage is:

$$(V_o)_{peak} = \omega R_F C V_m \qquad \text{(Eq. 3-7)}$$

In addition to a 90° phase shift (for a sine-wave input), the output is also inverted since the input signal is applied to the op-amp's inverting input.

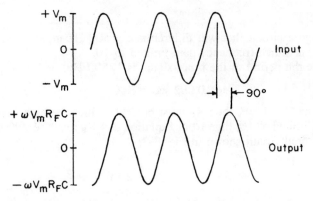

Fig. 3-4. Input and output waveforms for op-amp differentiator.

Example

What will be the peak output voltage for the circuit of Fig. 3-2 with $R_F = 200$ kΩ, $C = 0.01$ μF, given a 1-volt peak 200-Hz sine-wave input signal?

Using Equation 3-7, the peak output voltage is:

$$(V_o)_{peak} = \omega R_F C V_m$$
$$= (6.28)(200 \text{ Hz})(200 \text{ k}\Omega)(.01 \text{ }\mu\text{F})(1 \text{ V})$$
$$= 2.51 \text{ volts}$$

For the triangle-wave input signal shown in Fig. 3-5, the input frequency can be expressed as:

$$f = \frac{1}{t_1 + t_2} \qquad \text{(Eq. 3-8)}$$

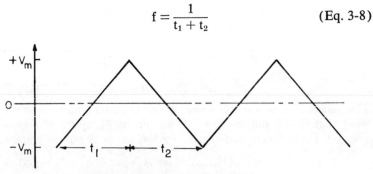

Fig. 3-5. Triangle-wave input signal.

During the time period t_1, the equation for the input signal, which is in the form of a straight line, is written as:

$$(V_i)_{t_1} = -V_m + 2\frac{V_m}{t_1}t \qquad \text{(Eq. 3-9)}$$

and for the time period t_2,

$$(V_i)_{t_2} = V_m - 2\frac{V_m}{t_2}t \qquad \text{(Eq. 3-10)}$$

Substituting Equations 3-9 and 3-10 into Equation 3-2 for the basic differentiator circuit, the output voltages for the two time periods are:

$$(V_o)_{t_1} = -R_F C\frac{d}{dt}\left[-V_m + 2\frac{V_m}{t_1}t\right]$$

$$= -R_F C(2V_m/t_1) \qquad \text{(Eq. 3-11)}$$

and,

$$(V_o)_{t_2} = R_F C(2V_m/t_2) \qquad \text{(Eq. 3-12)}$$

The output waveform will then be a square wave with a peak voltage equal to:

$$(V_o)_{peak} = \pm R_F C(2V_m/t_{1,2}) \qquad \text{(Eq. 3-13)}$$

During the time period t_1, the peak output will be negative, and during t_2, the peak output voltage will be positive, as shown in Fig. 3-6.

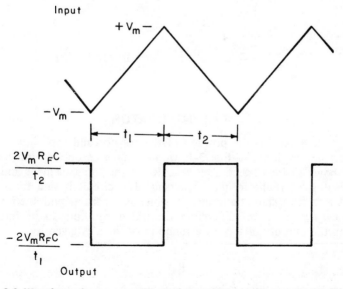

Fig. 3-6. Waveforms for op-amp differentiator with triangle-wave input signal.

Example

For the differentiator circuit of Fig. 3-2 with $R_F = 200$ kΩ and $C = 0.01$ μF, what will be the peak output voltage for a 100-Hz triangle-wave input whose peak voltage is 1 volt (i.e., 2 volts peak-to-peak), assuming that the triangle wave is symmetrical $(t_1 = t_2)$?

Since the time periods t_1 and t_2 are equal, then:

$$f = \frac{1}{t_1 + t_2} = 100 \text{ Hz}$$

or,

$$t_1 + t_2 = 0.01 \text{ sec}$$

so that,

$$t_1 = t_2 = 0.005 \text{ sec}$$

Using Equations 3-11 and 3-12,

$$(V_o)_{t_1} = -R_F C(2V_m/t_1)$$

$$= -\frac{(200 \text{ k}\Omega)(0.01 \text{ }\mu\text{F})(2)(1 \text{ V})}{(.005 \text{ sec})}$$

$$= -0.8 \text{ volt}$$

and,

$$(V_o)_{t_2} = +R_F C(2V_m/t_2)$$

$$= +\frac{(200 \text{ k}\Omega)(0.01 \text{ }\mu\text{F})(2)(1 \text{ V})}{(.005 \text{ sec})}$$

$$= +0.8 \text{ volt}$$

THE INTEGRATOR

By interchanging the position of the resistor and capacitor of the differentiator circuit of Fig. 3-1, we now have an op-amp integrator. As shown in Fig. 3-8, the resistor, R_1, is the input element and the capacitor, C, is the feedback element. The circuit is said to be the *inverse* of the differentiator circuit, which is consistent with the mathematical operations of differentiation and integration. In its integral form, the output voltage, as a function of time, is given by:

$$V_o = -\frac{1}{R_1 C} \int_0^t V_i \, dt \qquad \text{(Eq. 3-14)}$$

which represents "the area under the curve."

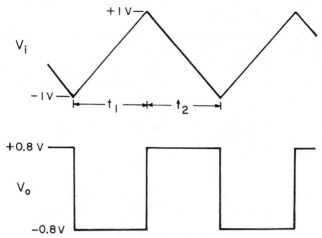

Fig. 3-7. Waveforms for op-amp differentiator with a 2-volt peak-to-peak triangle-wave input signal.

However, a more practical integrator circuit is shown in Fig. 3-9. The resistor, R_S, across the feedback capacitor, called a *shunt resistor*, is used to limit the *low-frequency gain* of the circuit. If the low-frequency gain were not limited, the dc offset, although small, would also be integrated over the integration period, eventually saturating

Fig. 3-8. The integrator.

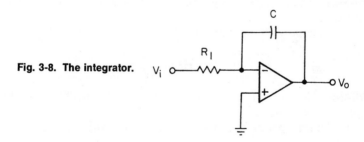

the op-amp. The dc offset voltage due to the input bias current is minimized by resistor R_2, which is equal to the parallel combination of the input and shunt resistors, or:

$$R_2 = \frac{R_1 R_S}{R_1 + R_S} \qquad \text{(Eq. 3-15)}$$

Since this shunt resistor helps to limit the circuit's low-frequency gain, Equation 3-14 is valid for input frequencies greater than:

$$f_c = \frac{1}{2\pi R_S C} \qquad \text{(Eq. 3-16)}$$

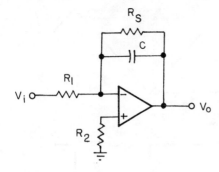

Fig. 3-9. Integrator with limited low-frequency gain.

For input frequencies less than f_c, the performance of Fig. 3-9 approaches that of an inverting amplifier with a voltage gain of:

$$\frac{V_o}{V_i} = -\frac{R_S}{R_1} \qquad \text{(Eq. 3-17)}$$

In practice, R_S is made approximately 10 times R_1. As with the differentiator circuit, the time constant R_1C is made approximately equal to the period of the input signal to be integrated.

Example

Using the square wave of Fig. 3-7 as the input signal for the integrator circuit of Fig. 3-9, determine R_1, R_S, and C, and the peak output voltage.

From the previous example, $f = 100$ Hz. Then if $C = 0.01$ μF,

$$100 \text{ Hz} = \frac{1}{R_1C}$$

Consequently, $R_1 = 1$ MΩ and $R_S = 10$ MΩ. From Equation 3-15, R_2 should be 910 kΩ.

For the time period t_1 (.005 sec), the output voltage is:

$$V_o = -\frac{1}{R_1C} \int_0^{t_1} V_i \, dt$$

$$= -\frac{1}{(1 \text{ M}\Omega)(.01 \ \mu\text{F})} \int_0^{t_1 = .005 \text{ sec}} (-0.8 \text{ V}) \, dt$$

$$= -(100)(-0.8) \int_0^{t_1 = .005 \text{ sec}}$$

$$= +0.4 \text{ volt}$$

In a similar manner, the output voltage for the time period t_2 is -0.4 volt.

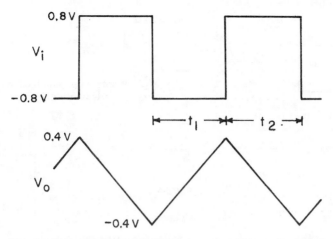

Fig. 3-10. Waveform for integrator with square-wave input signal.

AN INTRODUCTION TO THE EXPERIMENTS

The following two experiments are designed to demonstrate the design and operation of the op-amp differentiator and integrator at various frequencies. These two experiments that you will perform can be summarized as follows:

Experiment No.	Purpose
1	Demonstrates the design and operation of an op-amp differentiator.
2	Demonstrates the design and operation of an op-amp integrator.

EXPERIMENT NO. 1

Purpose

The purpose of this experiment is to demonstrate the design and operation of an op-amp differentiator, using a type 741 op-amp.

Schematic Diagram of Circuit (Fig. 3-11)

Design Basics

- Output voltage: $V_o = -R_F C \dfrac{dV_i}{dt}$

- Low-frequency response: $f_c = \dfrac{1}{2\pi R_S C}$

- When: $f < f_c$, the circuit acts as a differentiator,
  $f > f_c$, the circuit approaches that of an inverting amplifier with a voltage gain of $-R_F/R_S$.

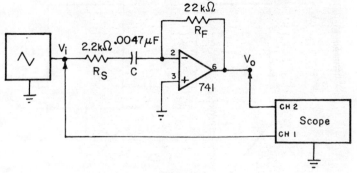

Fig. 3-11.

Step 1

Set your oscilloscope for the following settings:

- Channel 1: 0.5 volt/division
- Channel 2: 0.05 volt/division
- Time base: 0.5 msec/division
- DC coupling

Step 2

Wire the circuit shown in the schematic diagram and apply power to the breadboard.

Step 3

Adjust the peak-to-peak voltage of the triangle-wave input at 1 volt (0.5 volt peak) and the frequency at 400 Hz (2 complete cycles). As shown in Fig. 3-12, you should see that the output signal is a *square wave that is 180° out-of-phase with the input.*

Step 4

Temporarily disconnect the Channel 2 probe and note the position of the resulting straight line (0 volts). Reconnect the probe to the output of the circuit as shown in the schematic diagram (Fig. 3-11). Measure the negative peak voltage of the square wave (with respect to ground), recording your result below:

$$\text{Negative peak voltage} = \underline{\hspace{1cm}} \text{ volts}$$

Step 5

Now measure the time period for which the square-wave voltage is negative (t_1). If the input frequency is exactly 400 Hz and symmetrical, you should have measured approximately 1.25 msec.

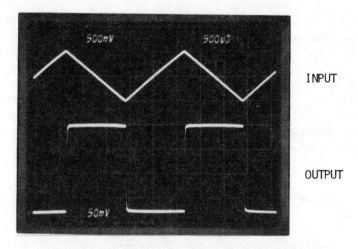

Fig. 3-12. Input and output waveforms for op-amp differentiator in Experiment No. 1.

Step 6

For this negative output voltage, the equation for the peak output voltage of a triangle-differentiated square wave is given by:

$$(V_o)_{peak} = -\frac{2R_F C V_m}{t_1}$$

How does the value that you determined in Step 4 compare with the predicted value?

When this experiment was performed, I measured a peak voltage of 0.0825 volt, as compared to a calculated value of 0.0827 volt.

Step 7

Change the time base to 0.2 msec/division and Channel 2 to 0.1 volt/division. Now adjust the input frequency so that there are 2 complete cycles (1 kHz). Repeat Steps 4, 5, and 6. How does your experimental result compare with the calculated value?

If the input frequency is exactly symmetrical and at 1 kHz, then t_1 is 0.5 msec, giving a calculated value for the peak output voltage of 0.205 volt. Your measured value should be near this value. At what approximate frequency will this circuit cease to act as a differentiator (that is, approach the operation of an inverting amplifier)?

From the low-frequency response formula, the demarcation between differentiator and amplifier operation is approximately 15.4 kHz.

Step 8

Now change the time base to 10 μsec/division and Channel 2 to 2 volts/division. Adjust the frequency so that there are 3 complete cycles shown on the oscilloscope's screen (30 kHz). What does the output signal look like?

The output signal should look like the input, but inverted, as shown in Fig. 3-13.

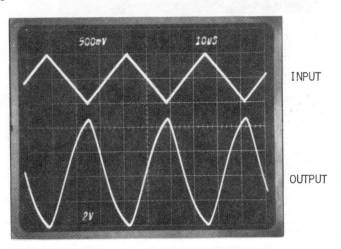

Fig. 3-13. Waveforms for differentiator circuit in Experiment No. 1 when it functions as an inverting amplifier.

At this frequency, the circuit functions as an inverting amplifier.

Step 9

Measure the peak-to-peak output voltage and determine the voltage gain. Is it what you would expect?

The voltage gain should be almost 10. In my experiment, I measured an output voltage of 9.6 volts peak-to-peak, giving a voltage gain of 9.6.

EXPERIMENT NO. 2

Purpose

The purpose of this experiment is to demonstrate the design and operation of an op-amp integrator, using a type 741 op-amp.

Schematic Diagram of Circuit (Fig. 3-14)

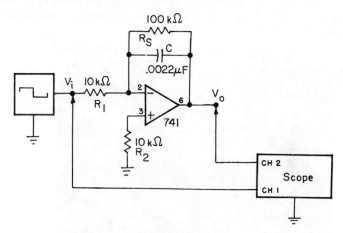

Fig. 3-14.

Design Basics

- Output voltage: $V_o = -\dfrac{1}{R_1 C} \displaystyle\int_0^t V_i \, dt$

- Low-frequency response: $f_c = \dfrac{1}{2\pi R_S C}$

- When: $f < f_c$, the circuit approaches that of an inverting amplifier with a voltage gain of $-R_S/R_1$.

  $f > f_c$, the circuit acts as an integrator.
- For minimum output offset due to input bias currents,

$$R_2 = \frac{R_1 R_S}{R_1 + R_S}$$

Step 1

Set your oscilloscope for the following settings:

- Channels 1 & 2: 0.5 volt/division
- Time Base: 20 μsec/division
- DC coupling

Step 2

Wire the circuit shown in the schematic diagram (Fig. 3-14), and apply power to the breadboard.

Step 3

Adjust the peak-to-peak voltage of the input square wave at 1 volt (0.5 volt peak) and the frequency at 10 kHz (2 cycles). As shown in Fig. 3-15, you should see that the output signal is a triangle wave that is 180° out-of-phase with the input.

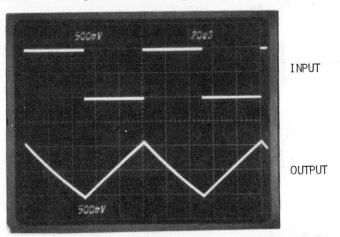

Fig. 3-15. Input and output waveforms for op-amp integrator in Experiment No. 2.

Step 4

Temporarily disconnect the Channel 2 probe and note the position of the resulting straight line (0 volts). Reconnect the probe to the output of the circuit as shown in the schematic diagram (Fig. 3-14). Measure the peak-to-peak triangle voltage, recording your result below:

Peak-to-peak voltage = _____ volts

Step 5

Now measure the time period that the input square-wave takes to complete one-half cycle (t). If the input is symmetrical and exactly 10 kHz, you should have measured 50 μsec.

Step 6

For a square-wave input signal, the output peak-to-peak voltage of the triangle wave is:

$$(V_o)_{p-p} = -\frac{V_m t}{R_1 C}$$

How does the value you determined in Step 4 compare with the value computed from the above formula?

Step 7

Change the oscilloscope's time base to 50 μsec/division and Channel 2 to 1 volt/division. Now adjust the input frequency so that there are 2 complete cycles (4 kHz). Repeat Steps 4, 5, and 6. How does your experimental result compare with the calculated value?

When I performed this step, I measured a peak-to-peak voltage of 2.70 volts over a 125-μsec time period. The calculated value was 2.84 volts. At what approximate frequency will this circuit cease to act as an integrator (that is, approach the operation of an inverting amplifier)?

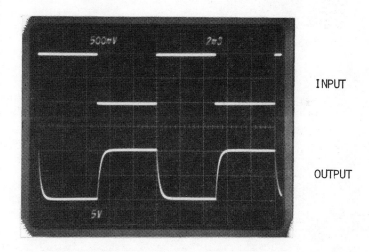

INPUT

OUTPUT

Fig. 3-16. Waveforms for integrator circuit in Experiment No. 2 when it functions as an inverting amplifier.

From the low-frequency response formula, the demarcation between integrator and amplifier operation is approximately 724 Hz.

Step 8

Now change the time base to 2 msec/division and Channel 2 to 5 volts/division. Adjust the frequency so that there are 2 complete cycles shown on the oscilloscope's screen (100 Hz). What does the output signal look like?

The output looks something like the input, but inverted, as shown in Fig. 3-16. At this frequency, the circuit nearly functions as an inverting amplifier.

Step 9

Measure the peak-to-peak output voltage and determine the voltage gain. Is it what you would expect?

The voltage gain should be approximately 10, since the ratio R_S/R_1 is 10.

CHAPTER 4

Voltage and Current Circuits

INTRODUCTION

In this chapter we will discuss some basic op-amp circuits whose input or output signal is a current, and which are useful in the formulation of current sources and metering circuits. In addition, current-to-voltage, and voltage-to-current converters are also possible.

OBJECTIVES

At the completion of this chapter, you will be able to do the following:

• Design and predict the performance of the following
    a constant-current source
    a current-to-voltage converter
    an inverting voltage-to-current converter
    a noninverting voltage-to-current converter

CONSTANT-CURRENT SOURCE

Using the circuit of Fig. 4-1, an op-amp can be made to function as a constant-current source. Here, the input voltage, such as a battery or other stable reference voltage, delivers a constant current through the input resistor R_1, which in turn must also flow through the feedback resistor R_2 (in an ideal op-amp, the input current is zero), or the load. By Ohm's law, the current through R_1, and, consequently, R_2 must be:

$$I = \frac{V_{REF}}{R_1} \qquad \text{(Eq. 4-1)}$$

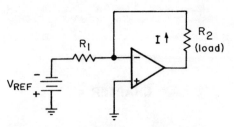

Fig. 4-1. Op-amp constant-current source.

If for some reason, the load resistance changes, *the current that flows through the load remains the same,* provided that V_{REF} and R_1 remain constant.

Current-to-Voltage Converter

The basic current-to-voltage converter shown in Fig. 4-2 is essentially an inverting amplifier, but without an input resistor. The in-

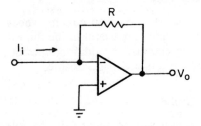

Fig. 4-2. Current-to-voltage converter.

put current I_i is applied directly to the inverting input of the op-amp. Since this input current also flows through the feedback resistor R (similar to the constant-current source), the output voltage is simply:

$$V_o = I_i R \qquad \text{(Eq. 4-2)}$$

For such a circuit, the op-amp's input bias current (I_b) is also added to the input current, so that Equation 4-2 is rewritten as:

$$V_o = (I_i + I_b)R \qquad \text{(Eq. 4-3)}$$

Consequently, care should then be taken to keep the input bias current small compared with the input current.

Example

An example of the current-to-voltage converter is the photocell circuit of Fig. 4-3.

Suppose that under fully lighted conditions, the current through the photocell is 100 μA, and when the photocell is fully covered from light, the current is 10 μA. Being only concerned with the change

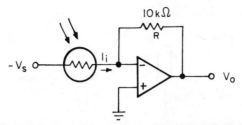

Fig. 4-3. Current-to-voltage converter employing a photocell.

in the output voltage, the input current changes 90 μA. Then according to Equation 4-2, the corresponding change in output voltage is:

$$\Delta V_o = \Delta I_i R$$
$$= (90 \ \mu A)(10 \ k\Omega)$$
$$= 0.9 \ volt$$

VOLTAGE-TO-CURRENT CONVERTERS

For applications such as driving relays and analog meters, a voltage-to-current converter, also called a *transmittance amplifier,* is frequently used. Depending on the particular application, a voltage-to-current converter can drive either floating or grounded loads.

For *floating* loads, the circuits of Figs. 4-4 and 4-5 can be used. Fig. 4-4 is an *inverting voltage-to-current converter,* and is similar in form to the inverting amplifier except that the feedback element

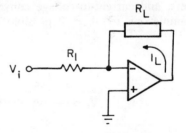

Fig. 4-4. Inverting voltage-to-current converter.

(the load) is now the coil of a relay, or possibly an ammeter with an internal resistance of R_L. Also, this circuit resembles the constant-current source of Fig. 4-1. The current flowing through the floating load is then:

$$I_L = \frac{V_i}{R_1} \qquad \text{(Eq. 4-4)}$$

and *this current is independent of the value of the load resistance R_L.*

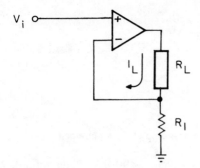

Fig. 4-5. Noninverting voltage-to-current converter.

Fig. 4-5 is a *noninverting voltage-to-current converter*. The load current for this circuit is the same as was given in Equation 4-4.

For loads that are grounded on one side, the circuit of Fig. 4-6 is used. Controlled by the input voltage V_i, the load current is given by:

$$I_L = -\frac{V_i}{R_3} \qquad \text{(Eq. 4-5)}$$

when,

$$\frac{R_4}{R_3} = \frac{R_2}{R_1} \qquad \text{(Eq. 4-6)}$$

AN INTRODUCTION TO THE EXPERIMENTS

The following experiments are designed to demonstrate the use of the op-amp in forming current sources, voltage-to-current converters, and current-to-voltage converters. In all these experiments, an ammeter is required. If possible, this meter should be a digital type,

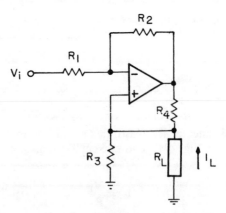

Fig. 4-6. Voltage-to-current converter for grounded loads.

giving more accurate results than can be obtained with an analog-type meter.

The experiments that you will perform can be summarized as follows:

Experiment No.	Purpose
1	Demonstrates the design and operation of a constant-current source, using a photocell as a variable load.
2	Demonstrates the design and operation of a current-to-voltage converter by measuring the current through a photocell.
3	Demonstrates the design and operation of a noninverting voltage-to-current converter.
4	Demonstrates the design and operation of an inverting voltage-to-current converter.

EXPERIMENT NO. 1

Purpose

The purpose of this experiment is to demonstrate the operation of a constant-current source, using a photocell as a variable load and a type 741 op-amp.

Schematic Diagram of Circuit (Fig. 4-7)

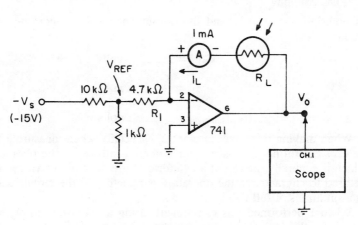

Fig. 4-7.

Design Basics

- $I_L = \dfrac{V_{REF}}{R_1}$, independent of R_L

- $V_o = -\dfrac{R_L}{R_1} V_{REF}$

Step 1

For this experiment you will need an ammeter with a full-scale reading of about 1 mA. A digital multimeter capable of measuring current is preferred for accurate results; however, a conventional analog meter can be used, but will have reduced accuracy. The photocell is a general-purpose type. The ones I used were obtained from Poly Paks, selling for about 50¢ each. Depending on the exact type you use, the experimental values that you will obtain may vary slightly.

Step 2

Wire the circuit shown in the schematic diagram (Fig. 4-7). Apply power to the breadboard and observe the ammeter's reading. If the polarity of the indicated current is *negative* (or the analog meter's needle is "pegged" to the left), *disconnect the power momentarily and reverse the meter's connections.*

Step 3

Now measure voltage V_{REF}, the reference voltage for the constant-current source, using an oscilloscope with the following settings:

- Channel 1: 0.2 volt/division
- Time base: 0.1 msec/division
- DC coupling

In addition, measure I_L and the output voltage V_o, recording your results below:

$$V_{REF} = \underline{\hspace{1cm}} \text{ volts}$$

$$I_L = \underline{\hspace{1cm}} \text{ mA}$$

$$(V_o)_{\text{full light}} = \underline{\hspace{1cm}} \text{ volts}$$

When making these measurements, especially when measuring V_o, be very careful not to obstruct the detecting surface of the photocell, which is generally the top of a cylindrical case. The above values will be used to characterize the operating parameters of the circuit under the conditions of full lighting.

When I performed this experiment, using a ±15-volt supply, my results for this step, as a comparison, were:

$$V_{REF} = -1.10 \text{ volts}$$
$$I_L = +0.230 \text{ mA}$$
$$(V_o)_{\text{full light}} = +0.52 \text{ volt}$$

Step 4

Since the photocell is a light-sensitive resistor, this circuit is identical to an *inverting amplifier*. From the design equation:

$$V_o = -\frac{R_L}{R_1}V_{REF}$$

determine the photocell's resistance (R_L) for the measurements in Step 3, recording it below:

$$(R_L)_{\text{full light}} = \underline{\hspace{1.5cm}} \Omega$$

For my experimental setup, this resistance was 2.22 kΩ.

Step 5

Now compare the measured current in Step 3 with the value calculated from:

$$I_L = \frac{V_{REF}}{R_1}$$

$$= \underline{\hspace{1.5cm}} \text{ mA}$$

Are the two values similar?

Depending on the exact value of R_1, these two values should almost be equal. In my case, I calculated a value of 0.234 mA, as compared with a measured value of 0.230 mA.

Step 6

Now monitor the ammeter and the output voltage on the oscilloscope while slowly passing your hand over the photocell. What happens?

You should have observed that *the output increases* when your hand is covering the photocell. However, *the current remains constant,* and is the same value that you measured in Step 3.

Step 7

Now place your finger directly on top of the photocell's surface. Measure I_L and V_o, recording your results below:

$$I_L = \underline{\hspace{2cm}} \text{ mA}$$

$$(V_o)_{dark} = \underline{\hspace{2cm}} \text{ volts}$$

Step 8

As in Step 4, determine the photocell's resistance for the situation when your finger covers the photocell's face, recording its value below:

$$(R_L)_{dark} = \underline{\hspace{2cm}} \Omega$$

In my experiment, I measured an output voltage of $+2.5$ volts, giving a "dark" resistance of 10.7 kΩ.

The resistance should change in Steps 4 and 7 (unless there is something wrong with your photocell), although the current stays the same. This is precisely the function of a constant-current source: *The output current is independent of the load.*

EXPERIMENT NO. 2

Purpose

The purpose of this experiment is to demonstrate the operation of a current-to-voltage converter, using a photocell and a type 741 op-amp.

Schematic Diagram of Circuit (Fig. 4-8)

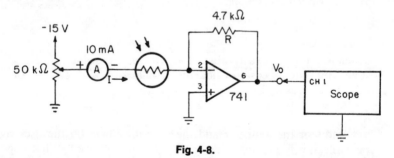

Fig. 4-8.

Design Basics

- $V_o = -IR$

Step 1

Set your oscilloscope for the following settings:

- Channel 1: 1 volt/division
- Time base: 0.1 msec/division
- DC coupling

Step 2

Wire the circuit shown in the schematic diagram (Fig. 4-8). Apply power to the breadboard. *If your ammeter indicates a negative value, momentarily disconnect the power and reverse the meter's leads.*

Step 3

Adjust the 50-kΩ potentiometer until the oscilloscope reads some convenient value for the output voltage, V_o, which should be less than the supply voltage. For my setup, I set the output voltage at +7.0 volts, using a 15-volt supply. Note the ammeter reading and record its value along with V_o below:

$$I_1 = \underline{\hspace{1.5cm}} \text{ mA}$$

$$(V_o)_1 = \underline{\hspace{1.5cm}} \text{ volts}$$

These will be the circuit's operating values under full-light conditions.

Step 4

Now cover the face of the photocell with your hand and again measure the current and output voltage, recording your results below:

$$I_2 = \underline{\hspace{1.5cm}} \text{ mA}$$

$$(V_o)_2 = \underline{\hspace{1.5cm}} \text{ volts}$$

These will be the circuit's operating values under dark conditions.

Step 5

Subtract I_2 from I_1, and $(V_o)_2$ from $(V_o)_1$

$$\Delta I = I_1 - I_2 = \underline{\hspace{1.5cm}} \text{ mA}$$

$$\Delta V_o = (V_o)_1 - (V_o)_2 = \underline{\hspace{1.5cm}} \text{ volts}$$

and divide ΔV_o by ΔI, so that:

$$\frac{\Delta V_o}{\Delta I} = R$$

$$= \underline{\hspace{1.5cm}} \Omega$$

How does this resistance value compare with the value for R in the circuit (i.e., the 4.7-kΩ feedback resistor)?

Depending on the exact value of the resistor you used, these two values should nearly be the same. In my setup, my results, for comparison, are:

$(V_o)_1$	$(V_o)_2$	I_1	I_2
7.0	5.4	1.428	1.098
$\Delta V_o = 1.6$ volts		$\Delta I = 0.33$ mA	

Consequently, $R = 4.85$ kΩ

The resistor R then transforms an input current I into a corresponding output voltage V_o, hence the name: current-to-voltage converter.

Step 6 (Optional)

Change R to 1 kΩ and repeat this experiment.

EXPERIMENT NO. 3

Purpose

The purpose of this experiment is to demonstrate the operation of a noninverting voltage-to-current converter, using a type 741 op-amp.

Schematic Diagram of Circuit (Fig. 4-9)

Design Basics

- $I_L = \dfrac{V_i}{R_1}$ (independent of R_L)

Step 1

Set your oscilloscope for the following settings:

- Channel 1: 0.5 volt/division
- Time base: 0.1 msec/division
- DC coupling

Step 2

Wire the circuit shown in the schematic diagram (Fig. 4-9). Initially, replace the load resistor R_L with a short circuit by connecting a wire across it. Apply power to the breadboard and adjust the potentiometer so that V_i equals +0.5 volt. If the ammeter reading is negative, momentarily disconnect the power and reverse the meter leads.

Step 3

With $V_i = +0.5$ volt, measure I_L and record your result below:

$$V_i = \underline{\quad +0.5 \quad} \text{ volt}$$

$$I_L = \underline{\hspace{2cm}} \text{ mA}$$

From the design equation, calculate the expected value for I_L. Is it the same as the measured value?

These two values should approximately be the same, which is 0.5 mA.

Step 4

Now remove the jumper wire across the 1-kΩ resistor. Did the load current I_L change?

You should have measured *no change in load current.*

Step 5

With the following combinations of input voltage (V_i) and load resistance (R_L), complete the following table:

V_i	R_L	I_L (measured)	I_L (calculated)
0.5 V	1 kΩ		
	10 kΩ		
	22 kΩ		
	27 kΩ		
	33 kΩ		
1.0 V	470 Ω		
	1 kΩ		
	3.3 kΩ		
	5.6 kΩ		
	10 kΩ		
	12 kΩ		
3.0 V	470 Ω		
	1 kΩ		
	3.3 kΩ		
	3.9 kΩ		

Step 6

From your experimental values in Step 5, what do you conclude about the operation of a voltage-to-current converter?

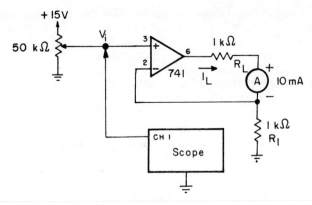

Fig. 4-9.

You should conclude that the current produced is independent of the load resistance, being dependent only upon the input voltage and the input resistor R_1.

Step 7

Did you observe different ammeter readings for the final resistance value for each input voltage (e.g., $V_i = 0.5$ V, $R_L = 33$ kΩ, $V_i = 1$ V, $R_L = 12$ kΩ, etc.) compared with the other values of load resistance within each group?

In most cases, you may have observed that the load current is less for a lower load resistance for the same input voltage. Although the current is basically independent of the load resistance, there is nevertheless a maximum value of R_L that can be used.

You also should have noticed that this circuit is identical to *a noninverting amplifier,* so that the output voltage is then:

$$V_o = \left[1 + \frac{R_L}{R_i} \right] V_i$$

For the condition, $V_i = 0.5$ volt, and $R_L = 33$ kΩ, we find that the output voltage is:

$$V_o = \left[1 + \frac{33 \text{ k}\Omega}{1 \text{ k}\Omega} \right] (0.5 \text{ volt})$$

$$= (34)(0.5 \text{ volt})$$

$$= 17.0 \text{ volts}$$

which is greater than the op-amp's supply voltage. Using a ±15-volt supply, the op-amp's output voltage will saturate at approximately

13 volts, thus reducing the maximum current that can be supplied to the load. Also, it is impossible for the op-amp to deliver an output voltage that is greater than its supply voltage.

If on the other hand, a ±6-volt supply is used, you would find that the maximum load resistance will be even lower, following the same reasoning as above.

EXPERIMENT NO. 4

Purpose

The purpose of this experiment is to demonstrate the operation of an inverting voltage-to-current converter, using a type 741 op-amp. This experiment is similar to Experiment No. 3.

Schematic Diagram of Circuit (Fig. 4-10)

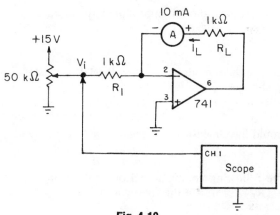

Fig. 4-10.

Design Basics

- $I_L = \dfrac{V_i}{R_1}$ (independent of R_L)

Step 1

Set your oscilloscope for the following settings:

- Channel 1: 0.5 volt/division
- Time base: 0.1 msec/division
- DC coupling

Step 2

Wire the circuit shown in the schematic diagram (Fig. 4-10). Initially, replace the load resistor R_L with a short circuit, by connecting a wire across it. Apply power to the breadboard and adjust the potentiometer so that V_i equals $+0.5$ volt. If the ammeter reading is negative, momentarily disconnect the power and reverse the meter leads.

Step 3

With $V_i = +0.5$ volt, measure I_L and record your result below:

$$V_i = \quad +0.5 \quad \text{volt}$$

$$I_L = \underline{\hspace{2cm}} \text{ mA}$$

From the design equation, calculate the expected value for I_L. Is it the same as the measured value?

These two values should approximately be the same, which is 0.5 mA.

Step 4

Now remove the jumper wire across the 1-kΩ resistor. Did the load current I_L change?

You should have measured *no change in load current*.

Step 5

With the following combinations of input voltage (V_i) and load resistance (R_L), complete the following table:

V_i	R_L	I_L (measured)	I_L (calculated)
0.5 V	1 kΩ		
	10 kΩ		
	22 kΩ		
	27 kΩ		
	33 kΩ		
1.0 V	470 Ω		
	1 kΩ		
	10 kΩ		
	12 kΩ		

V_I	R_L	I_L (measured)	I_L (calculated)
1.0 V	15 kΩ		
3.0 V	470 Ω		
	1 kΩ		
	3.3 kΩ		
	3.9 kΩ		
	4.7 kΩ		

Step 6

From your experimental values in Step 5, what do you conclude about the operation of a voltage-to-current converter?

You should conclude that the current produced is independent of the load resistance, being dependent only upon the input voltage and the input resistor R_1.

Step 7

Did you observe different ammeter readings for the final resistance value for each input voltage, compared with the other values of load resistance within each group?

In most cases, you may have observed that the load current is less for a lower resistance for the same input voltage. Although the current is basically independent of the load resistance, there is nevertheless a maximum value of R_L that can be used. In addition, you should have noticed that this circuit is identical to an *inverting amplifier,* and the same reasoning applies, as was done in the previous experiment.

CHAPTER 5

Nonlinear Signal Processing Circuits

INTRODUCTION

In this chapter, we shall present several nonlinear circuits using op-amps, which include those situations for which the output is essentially not a sine wave, or those when the op-amp's output approaches its maximum positive or negative excursion.

OBJECTIVES

At the completion of this chapter, you will be able to do the following:

- Design and predict the performance of the following:
  a comparator
  a peak detector
  a precision half-wave rectifier
  a precision full-wave rectifier
  a logarithmic amplifier
- Explain how to multiply and divide two input signals.

THE COMPARATOR

A comparator is a circuit that compares an input voltage with a reference voltage. The output of the comparator then indicates whether the input signal is either above or below the reference voltage. As shown for the basic circuit in Fig. 5-1, the output voltage approaches the positive supply voltage when the input signal is slightly greater

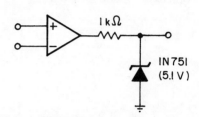

Fig. 5-1. The comparator.

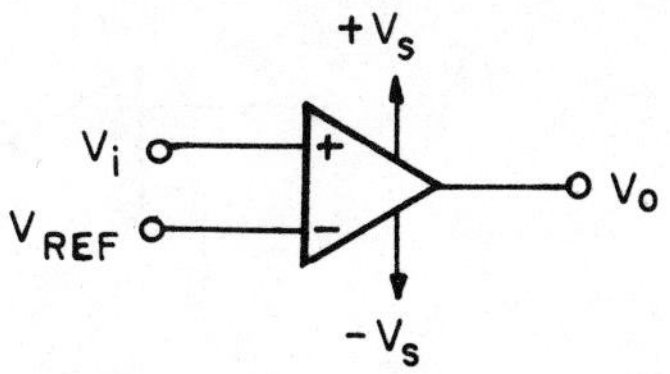

than the reference voltage, V_{REF}. When the input is slightly less than the reference, the op-amp's output approaches the negative supply voltage. Consequently, the exact threshold is dominated by the op-amp's input offset voltage, which should be nulled out.

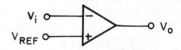

Fig. 5-2. Comparator limiting with a zener diode.

If the output voltage of the comparator is larger than required for a given application, such an interfacing with +5-volt TTL integrated circuits, the output can be limited by a suitable zener diode, as shown in Fig. 5-2 for TTL IC's.

Fig. 5-3. An inverting comparator.

When using op-amps as a comparator, the op-amp used should have a fast slew rate if it is to switch from one state to the other. Since external compensation reduces the op-amp's slew rate, it is best to use an uncompensated op-amp such as the type 318, having a slew rate of 70 V/μsec.

The circuits of Figs. 5-1 and 5-2 are *noninverting comparators,* so that the output voltage has the same polarity as its input. By reversing the inputs, as shown in Fig. 5-3, we then have an *inverting comparator.*

One nice application for the comparator is that it can convert a sine wave into a square wave, using the circuit of Fig. 5-4, so that the noninverting input is grounded (i.e., $V_{REF} = 0$).

By combining both a noninverting and an inverting comparator, both having different reference voltages, we can form a *window comparator,* which detects whether or not an input voltage V_i is between

111

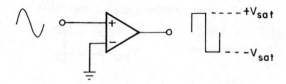

Fig. 5-4. Sine-wave to square-wave converter.

the limits V_L and V_H, the ("window"). As shown in Fig. 5-5, the comparators' outputs are logically combined by the two diodes. When the input voltage is between V_L and V_H, the output voltage is zero; otherwise, it equals $+V_{SAT}$.

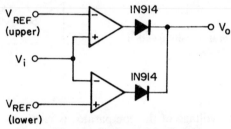

Fig. 5-5. A window comparator.

With the circuits discussed in this section, we can use the LED circuit of Fig. 5-6 to indicate that the comparator's output is positive or negative.* When the LED is lit, the output is positive.

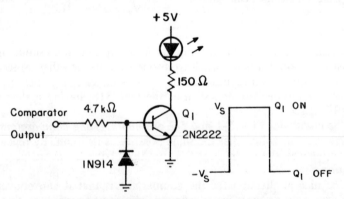

Fig. 5-6. Comparator circuit with LED indicator.

---

*Jung, W. G. *IC Op-Amp Cookbook*. Howard W. Sams & Co., Inc., Indianapolis, 1974, p. 235.

112

THE PEAK DETECTOR

The peak detector is a circuit that "remembers" the peak value of a signal. As shown in Fig. 5-7, when a *positive voltage* is fed to the noninverting input after the capacitor has been momentarily shorted (reset), the output voltage of the op-amp forward biases the diode and charges up the capacitor. This charging lasts until the inverting

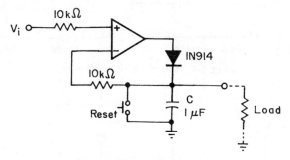

Fig. 5-7. The peak detector.

and noninverting inputs are at the same voltage, which is equal to the input voltage. When the noninverting input voltage exceeds the voltage at the inverting input, which is also the voltage across the capacitor, the capacitor will charge up to the new peak value. Consequently, the capacitor voltage will always be *equal to the greatest positive voltage applied to the noninverting input.*

Once charged, the time that the peak detector "remembers" this peak value is typically several minutes and depends on the impedance of the load that is connected to the circuit. Consequently, the capacitor will slowly discharge toward zero. To minimize this rate of discharge, a voltage follower can be used to buffer the detector's output from any external load, as shown in Fig. 5-8. Momentarily shorting the capacitor to ground immediately sets the output to zero.

"Precision" Rectifiers

When a diode is used as a rectifier to change an ac signal to a pulsating dc signal, the diode does not begin to conduct until the voltage drop across the diode is greater than 0.3 volt (for germanium types) or 0.7 volt (for silicon types). Consequently, diodes by themselves are not suitable for small-signal rectification.

The *half-wave rectifier,* shown in Fig. 5-9, will rectify small input signals. When the input signal is positive, all the current in the feedback loop flows through D_1 and the output voltage of the circuit will be zero. When the input signal is negative, the current in the feedback loop flows through diodes D_1 and D_2, so that the input voltage,

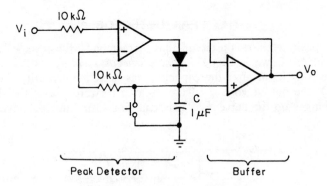

Fig. 5-8. Peak detector with buffer.

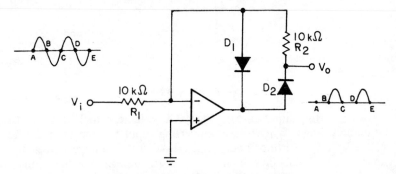

Fig. 5-9. "Precision" half-wave rectifier.

which appears inverted across R_2, also is the output voltage. Since the op-amp has high gain, a very small negative-going input is sufficient to make D_2 conduct. For this reason, this circuit is commonly referred to as a "precision" half-wave rectifier.

A *full-wave* "precision" rectifier is formed by summing the input and output voltages of the half-wave rectifier, as shown in Fig. 5-10.

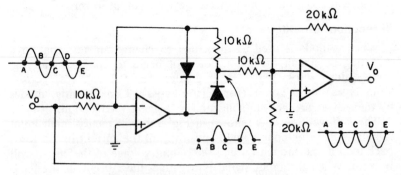

Fig. 5-10. "Precision" full-wave rectifier.

LOGARITHMIC AMPLIFIERS

A logarithmic amplifier has an output voltage that is proportional to the *logarithm* of the input, or:

$$V_o \propto \log V_i \qquad \text{(Eq. 5-1)}$$

For a logarithmic amplifier to function properly, its nonlinear element, such as a diode or transistor, must have logarithmic function. For a diode, the voltage drop across it (V_D) as a function of the current that flows through it is essentially given by the relation:

$$V_D = A \log (I) \qquad \text{(Eq. 5-2)}$$

where the constant A is based on the semiconductor properties of the diode.

For building practical logarithmic amplifiers, transistors are usually preferred over diodes, as shown in the transdiode logarithmic amplifier circuit of Fig. 5-11 using a grounded base npn transistor in the feedback loop when the input is positive.

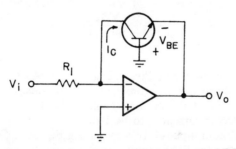

Fig. 5-11. The transdiode logarithmic amplifier.

The transdiode logarithmic amplifier uses Equation 5-2 as its basis, where the diode voltage drop is the base-to-emitter junction voltage of the transistor, and the current is the transistor's collector current, so that:

$$V_{BE} = A \log (I_C) \qquad \text{(Eq. 5-3)}$$

From the circuit of Fig. 5-11, we evolve to the more practical circuit of Fig. 5-12*. The capacitor across the npn transistor is used to reduce the ac gain while the diode protects the transistor against excessive reverse base-to-emitter voltage. In general, resistor R_1 is determined by the inequality pair:

* Lenk, J. D. *Manual for Operational Amplifier Users.* Reston Publishing Co., Inc., Reston, 1976, pp. 193-7.

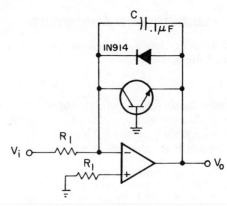

Fig. 5-12. Improved logarithmic amplifier.

$$R_1 \geq \frac{(V_i)_{max}}{(I_C)_{max}} \qquad \text{(Eq. 5-4)}$$

and,

$$R_1 \leq \frac{(V_i)_{min}}{\text{input bias current of op-amp}} \qquad \text{(Eq. 5-5)}$$

Example

Design a logarithmic amplifier, using the circuit of Fig. 5-12, having an input voltage varying from 1 mV to 10 volts. Assume that the input bias current of the op-amp (e.g., a 741) is 80 nA, and the maximum collector current is to be 1 mA.

With a maximum input of 10 volts, using Equation 5-4, we obtain a minimum value for R_1, so that:

$$R_1 \geq \frac{10 \text{ V}}{1 \text{ mA}}$$

$$\geq 10 \text{ k}\Omega$$

Using Equation 5-5, we determine the maximum value for R_1, so that:

$$R_1 \leq \frac{1 \text{ mV}}{80 \text{ nA}}$$

$$\leq 12.5 \text{ k}\Omega$$

Therefore, R_1 should be between 10 kΩ and 12.5 kΩ.

Logarithmic amplifiers must have provisions for cancelling the small dc input offset voltage, since it will also be logarithmically converted. If you use an op-amp, such as a type 709, that does not provide for external offset as does the 741, the circuit of Fig. 5-13 can be used for the previous example.

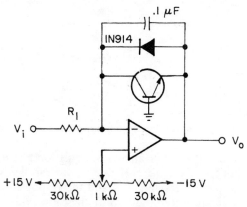

Fig. 5-13. Offset circuit for logarithmic amplifiers.

By interchanging the position of the input and feedback elements of the basic logarithmic circuit of Fig. 5-12, we have an *antilogarithmic* amplifier, or *inverse-log amplifier,* as shown in Fig. 5-14. Using both the logarithmic and antilogarithmic amplifier circuits, we can either multiply or divide input voltages.

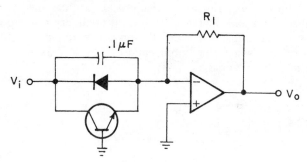

Fig. 5-14. The antilogarithmic amplifier.

Recalling from algebra, two basic relationships of logarithms are:

$$\log (AB) = \log A + \log B \qquad \text{(Eq. 5-6)}$$

and,

$$\log (A/B) = \log A - \log B \qquad \text{(Eq. 5-7)}$$

By summing the logarithms of two input voltages A and B, and using the antilog circuit, we can then obtain the product of A and B, as shown in Fig. 5-15. Instead of using a summing amplifier, a difference amplifier (analog subtractor) can be used so that the circuit of Fig. 5-16 is used to take the quotient of A and B.

117

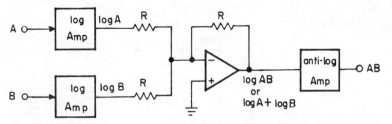

Fig. 5-15. Multiplication of two input signals using log and antilog amplifiers.

The circuits in this section all use a transistor to provide the required logarithmic characteristic. However, all transistors do not give adequate logarithmic characteristics, and the response is also subject to ambient temperature variations. For these and other reasons beyond the scope of this book,* you should strongly consider the use of commercial logarithmic modules, such as the 755, which is manufactured by Analog Devices, and is specifically designed for this purpose.

AN INTRODUCTION TO THE EXPERIMENTS

The following experiments in this chapter are designed to demonstrate the design and operation of several nonlinear op-amp circuits. These are applications for which the op-amp's output signal is not a sine wave, or, when the output approaches it maximum positive and/or negative excursion.

The experiments that you will perform can be summarized as followes:

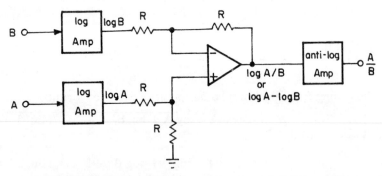

Fig. 5-16. Division of two input signals using log and antilog amplifiers.

---

* One good article on logarithmic amplifiers is: Sheingold, D., and Pouliot, F. "The Hows and Whys of Log Amps," Electronic Design, February 1, 1974, pp. 52-9.

EXPERIMENT NO. 1

Purpose

The purpose of this experiment is to demonstrate the operation of a simple noninverting comparator, using a type 741 op-amp and a LED indicator circuit.

Schematic Diagram of Circuit (Fig. 5-17)

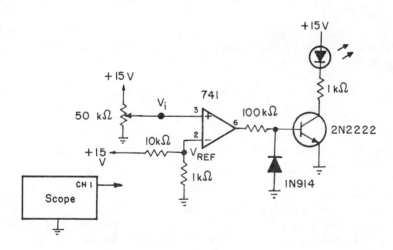

Fig. 5-17.

Design Basics

- When: $V_i < V_{REF}$, $V_o = -V_{SAT}$ (LED OFF)
- $V_i > V_{REF}$, $V_o = +V_{SAT}$ (LED ON)

119

Step 1

The following oscilloscope settings are dependent on the supply voltage that you will use. If a ±15-volt supply is used, set your scope to:

- Channel 1: 1 volt/division
- Time base: 1 msec/division
- DC coupling

If a ±6-volt supply is used, you will have to use a higher sensitivity for Channel 1, such as 0.5 volt/division.

Step 2

Wire the circuit shown in the schematic diagram (Fig. 5-17), then make sure that you have the npn transistor wired correctly, and the 1N914 diode connected correctly. When you are satisfied, then apply the power to the breadboard.

Step 3

Depending on the setting of the potentiometer, the LED may or may not be lit when you connect the power. If the LED is ON, turn the potentiometer past the point at which the LED is OFF.

Step 4

With the oscilloscope probe, measure the voltage at the junction of the 10-kΩ and 1-kΩ resistors at pin 2 of the 741 op-amp, and record your result below:

$$V_{REF} = \underline{\hspace{1cm}} \text{ volts}$$

This value is the reference voltage of the comparator, and will be dependent on the supply voltage that you are using. For the ±15-volt supply, this will be about 1.4 volts. If a ±6-volt supply is used, V_{REF} will be about 0.5 volt.

Step 5

Now connect the probe of the oscilloscope to pin 3 (the noninverting input) of the 741 op-amp. While watching the LED, vary the potentiometer just until the LED comes ON. Measure the voltage at pin 3, and record your result below:

$$(V_1)_{LED\ ON} = \underline{\hspace{1cm}} \text{ volts}$$

How does this value compare with the value you determined in Step 4?

You should find that these two values are about the same. When the input voltage (V_i) exceeds the reference voltage (V_{REF}) applied at the op-amp's inverting input, the output of the op-amp switches from its *negative saturation value* (usually about 1.4 volts less than the negative supply voltage) to the positive saturation voltage. With a ±15-volt supply, this corresponds to switching from about −13 to +13 volts. We have added this transistor and LED circuit to the comparator to visually determine whether or not the input voltage has exceeded the reference voltage.

Step 6

Disconnect the power to the breadboard. Verify the operation of the noninverting comparator by varying resistor R_1 and repeating Steps 3 through 5, according to the following table:

R_1	measured V_{REF}	measured $(V_i)_{LED\ ON}$
1 kΩ		
2.2 kΩ		
5.6 kΩ		

EXPERIMENT NO. 2

Purpose

The purpose of this experiment is to demonstrate the operation of an op-amp comparator as a sine-to-square wave converter, using a 741 op-amp.

Schematic Diagram of Circuit (Fig. 5-18)

Design Basics

- When: $V_i < 0$, $V_o = +V_{SAT}$
  $V_i > 0$, $V_o = -V_{SAT}$
  since $V_{REF} = 0$

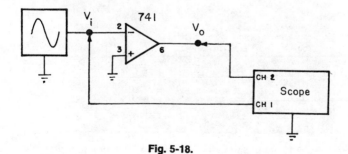

Fig. 5-18.

Step 1

Set your oscillopscope for the following settings:

- Channel 1: 1 volt/division
- Channel 2: 10 volts/division
- Time base: 1 msec/division
- DC coupling

Step 2

Wire the circuit shown in the schematic diagram (Fig. 5-18). Apply the power to the breadboard and adjust the input peak-to-peak voltage at 3 volts; also, adjust the input frequency so that there are about 3 complete cycles per the 10 horizontal divisions (300 Hz). What is the polarity of the output voltage when the input signal goe positive? When the input goes negative?

When the input signal is applied to the op-amp's inverting input the output signal's polarity will be the opposite of the input's. Th output will then be a square wave that is 180° out-of-phase with the sine-wave input. Consequently, this type of circuit is called an *inverting comparator*. Since the reference voltage is zero (the non inverting input is grounded), the output immediately goes negative when the input signal goes positive, as shown in Fig. 5-19.

Step 3

Disconnect the power to the breadboard. Then reverse the input connections to the op-amp, so that the sine-wave input signal is now

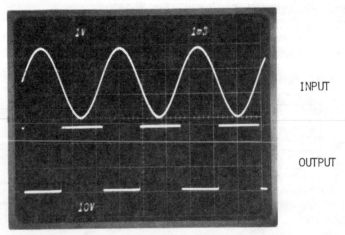

INPUT

OUTPUT

Fig. 5-19. Input and output waveforms for inverting comparator.

applied to the noninverting input while the inverting input is grounded.

Step 4

Reapply the power to the breadboard. The input peak-to-peak voltage is still 3 volts and at a frequency of 300 Hz. What is the difference between the operation of this comparator and the one you used in Steps 1 and 2?

For this comparator, the output square wave has the same polarity as the input sine-wave, as shown in Fig. 5-20. This circuit is then

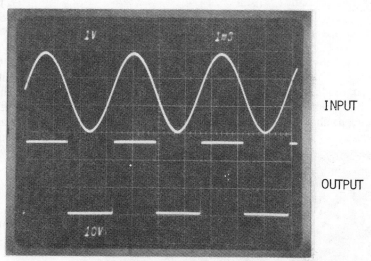

Fig. 5-20. Input and output waveforms for noninverting comparator.

called a *noninverting comparator. This circuit follows the definition of a noninverting amplifier, except that the output is now a square wave.*

Step 5

Again disconnect the power to the breadboard. Now add the circuit shown in Fig. 5-21 to the output of the noninverting comparator.

Fig. 5-21. Interface circuit for comparator outputs greater than 5.1 volts.

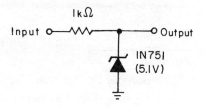

This type of interfacing is used when the output voltage of the comparator is larger than required for a given application. With this zener diode (5.1 volts) we are able to interface this comparator with +5-volt TTL integrated circuit chips.

Step 6

Apply the power to the breadboard. Then set your oscilloscope for the following settings:

- Channel 1: 0.5 volt/division
- Channel 2: 5 volts/division
- Time base: 0.2 msec/division
- DC coupling

Adjust the sine-wave input voltage at 1 volt peak-to-peak at a frequency of about 1500 Hz (3 full cycles per 10 horizontal divisions). You should notice that the output square wave has a peak-to-peak voltage of approximately 5 volts, which is equal to the rating of the zener diode used. By using a suitable zener diode, we are able to limit the output voltage of the comparator to almost any convenient value, as shown in Fig. 5-22 for a 5.1-volt zener diode.

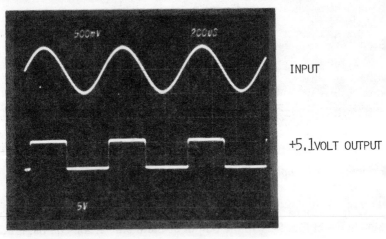

Fig. 5-22. Zener diode limits output voltage to 5.1 volts peak-to-peak.

EXPERIMENT NO. 3

Purpose

The purpose of this experiment is to demonstrate the design and operation of a window comparator, using two type 741 op-amps.

Schematic Diagram of Circuit (Fig. 5-23)

Design Basics

- When: $V_L < V_i < V_H$, $V_o = 0$
  otherwise: $V_o = +V_{SAT}$

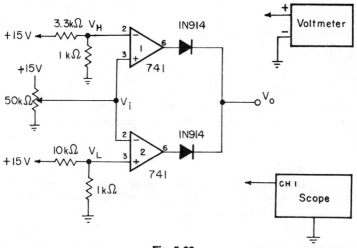

Fig. 5-23.

Step 1

The following oscilloscope settings are dependent on the supply voltage that is used. If a ±15-volt supply is used, set the oscilloscope to the following:

- Channel 1: 5.0 volt/division
- Time base: 1 msec/division
- DC coupling

Step 2

Wire the circuit shown in the schematic diagram (Fig. 5-23). Then apply the power to the breadboard.

Step 3

With the voltmeter (preferably a digital type or a vtvm) connected to pin 2 of the 1st op-amp, measure the voltage V_H, *the upper reference voltage of the window,* and record your result below:

$$V_H = \underline{\hspace{1cm}} \text{ volts}$$

Naturally, this value will be dependent on the supply voltage. Using a ±15-volt supply, this reference should be about 3.4 volts.

Step 4

Now measure the voltage at pin 3 of the 2nd op-amp, V_L, *the lower reference voltage of the window,* and record your result below:

$$V_L = \underline{} \text{ volts}$$

Again, this value will be dependent on the supply voltage. With a ±15-volt supply, this reference should be about 1.3 volts.

Step 5

Now connect the + probe of the voltmeter to the junction of pin 3 and 2 of the 1st and 2nd op-amps, respectively. Vary the 50-kΩ potentiometer so that the input voltage V_i is at its minimum value (almost 0 volts). As seen on the oscilloscope, what is the output voltage of the window comparator (V_o)?

The output voltage of the window comparator should equal $+V_{SAT}$. If you are using a ±15-volt supply, this will be approximately $+13$ volts. Otherwise, for other supply voltages, V_{SAT} will be approximately 1.4 volts less than the supply voltage. *When the input voltage of this window comparator is less than the lower reference V_L, the output voltage equals V_{SAT}.*

Step 6

Now slowly vary the potentiometer, increasing the input voltage V until the output voltage, as seen on the oscilloscope, suddenly drops to zero. Record the input voltage below:

$$V_{i_{(V_o = 0)}} = \underline{} \text{ volts}$$

How does this value compare with the voltage that you determined in Step 4?

These two voltages should be approximately the same. *The output of the window comparator will be zero when the input voltage exceeds the lower reference V_L.*

Step 7

Slowly increase the input voltage further just until the output voltage again is equal to V_{SAT} (i.e., the same output voltage you measured in Step 5). At this point, measure the input voltage with the voltmeter and record it below:

$$V_{i_{(V_o = +V_{SAT})}} = \underline{\hspace{2cm}} \text{ volts}$$

How does this value compare with the voltage that you determined in Step 3?

They should be approximately the same! The output of the window comparator again is equal to the saturated output voltage of the op-amp when the input voltage exceeds the comparator's upper reference voltage V_H. Consequently, we can now state the basic operation of this window comparator as follows:

The output voltage of this window comparator will be zero only when the input voltage is between the lower reference V_L and the upper reference V_H. Otherwise, if the input voltage is less than V_L, or is above V_H, the output equals the op-amp's saturated output voltage.

The "window" is the voltage range between the lower reference and the upper reference. It should be emphasized that there is nothing sacred about the numerical values for these two limits which you determined in Steps 3 and 4. These two values were picked at random with a couple of resistors that were within reach on the laboratory table.

Step 8

Verify the operation of the window comparator for a different set of voltage references. For example, change the 3.3-kΩ resistor to 1 kΩ and the 10-kΩ resistor to 4.7 kΩ, and repeat Steps 1 through 7.

EXPERIMENT NO. 4

Purpose

The purpose of this experiment is to demonstrate the design and operation of a "precision" half-wave rectifier, using a type 741 op-amp.

Schematic Diagram of Circuit (Fig. 5-24)

Design Basics

• Peak output voltage:

$$(V_o)_{peak} = 0, \quad \text{for } V_i > 0$$
$$= -\frac{R_2}{R_1}(V_{i\,peak}), \quad \text{for } V_i < 0$$

Step 1

Set your oscilloscope for the following settings:

- Channels 1 & 2: 2 volts/division
- Time base: 1 msec/division
- DC coupling

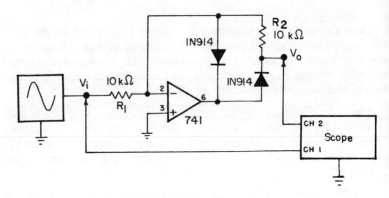

Fig. 5-24.

Step 2

Wire the circuit shown in the schematic diagram (Fig. 5-24). Apply power to the breadboard and adjust the input voltage at 4 volts peak-to-peak, and the frequency so that there are 3 complete cycles (300 Hz).

Step 3

On the oscilloscope (Channel 2), you should see a waveform whose output is zero when the input signal goes positive, but equal and opposite in polarity to the input when the input signal goes negative. Since the ratio R_2/R_1 is equal to 1, the peak-to-peak output voltage therefore equals the peak (*not* peak-to-peak) input voltage, as shown in Fig. 5-25.

Step 4

Change R_1 to 5 kΩ by placing two 10-kΩ resistors in parallel. What changes do you now see in the circuit's output voltage? Why?

The peak output voltage should now be equal to 4 volts, or twice the peak input voltage, since the ratio R_2/R_1 is now equal to 2.

Save this circuit as it will be used as part of the circuit in the next experiment.

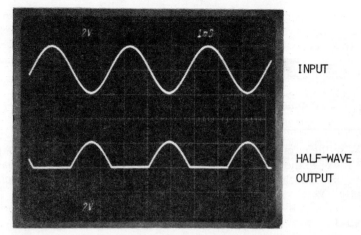

INPUT

HALF-WAVE
OUTPUT

Fig. 5-25. Input and output waveforms for precision half-wave rectifier.

EXPERIMENT NO. 5

Purpose

The purpose of this experiment is to demonstrate the design and operation of a "precision" full-wave rectifier, using two type 741 op-amps. The op-amp full-wave rectifier is formed by using the half-wave section of the previous experiment with a summing amplifier.

Schematic Diagram of Circuit (Fig. 5-26)

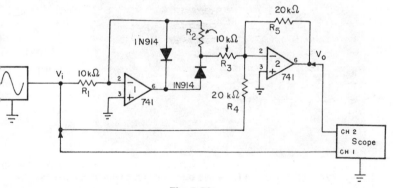

Fig. 5-26.

Design Basics

- Peak output voltage: $(V_o)_{peak} = - (V_i)_{peak}$
  when $R_1 = R_2 = R_3$
  $R_4 = R_5 = 2R_1$
- Output frequency: twice the input frequency

Step 1

Set your oscilloscope for the following settings:

- Channels 1 & 2: 2 volts/division
- Time base: 1 msec/division
- DC coupling

Step 2

Wire the circuit shown in the schematic diagram (Fig. 5-26). Apply power to the breadboard. Adjust the input voltage at 4 volts peak-to-peak, and the frequency so that there are 3 complete cycles (300 Hz).

Step 3

On the oscilloscope display (Channel 2), you should see a waveform whose shape exhibits a sinusoidal "hump" every time the input signal goes positive and negative with respect to ground. If we were to simultaneously display the input, half-wave and full-wave output signals, we would observe the waveform shown in Fig. 5-27.

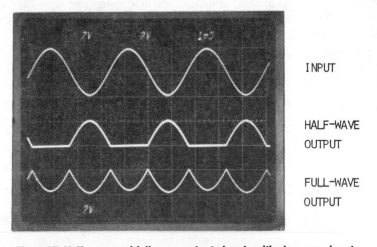

Fig. 5-27. Half-wave and full-wave output signals with sine-wave input.

When the input signal goes positive, the half-wave rectifier output is zero. When, at this point, both the input and half-wave signals are added together by the summing amplifier, the output of the 2nd op-amp is the same as the positive "hump" of the input, but of negative polarity. When the input signal goes negative, the half-wave rectifier output has a positive "hump." When this positive "hump" is added to the negative cycle of the input, the output is a "hump" similar to the previous half-cycle. The input signal is amplified by a factor of 1, while the half-wave signal is amplified by a factor of 2 (i.e., 20 kΩ/ 10 kΩ). As a result of this full-wave rectification, the output frequency is twice that of the input.

CHAPTER 6

Generators

INTRODUCTION

In this chapter we shall briefly discuss the application of op-amps to generate various kinds of simple waveforms. However, it should be kept in mind that there now are several integrated circuits that will perform the same function.

OBJECTIVES

At the completion of this chapter, you will be able to do the following:

• Design and build op-amp circuits to generate:

> sine and cosine waves
> triangle waves
> square waves
> staircase waves

SINE WAVES

The sine-wave oscillator, shown in Fig. 6-1, is called a *Wien-bridge* oscillator. The resistor-capacitor combinations R_1-C_1 and R_2-C_2 provide a *positive* feedback path around the op-amp, while resistor R_3 and the lamp L_1 provide *negative* feedback. It is the application of positive feedback that causes the circuit to oscillate with a sine-wave output. The frequency of oscillation is given by:

$$f_o = \frac{1}{2\pi R_1 C_1}$$

(Eq. 6-1)

where,

$$R_1 = R_2$$
$$C_1 = C_2$$

The lamp helps to regulate the amount of negative feedback, stabilizing the amplitude of the sine-wave output. Resistors R_3 and R_4 are used to account for lamp tolerances, so that this series combination approximately equals 750 Ω. As discussed in *IC Op-Amp Cookbook,* by Walter Jung, the value for R_1 depends on the type of op-amp used, which is summarized below in Table 6-1.

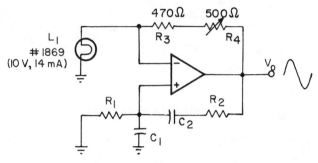

Fig. 6-1. Wien-bridge sine-wave oscillator.

Table 6-1. Value of R_1 and Operating Frequency for Various Op-Amps

R_1	Op-Amp
$\geq$ 1 MΩ	108, 1556, 8007
$\leq$ 1 MΩ	118, 107, 741
f_o	Op-Amp
$\geq$ 1 kHz	118, 1556, 8007
$\leq$ 1 kHz	107, 108, 741, 1556, 8007

Another commonly used sine-wave oscillator is the *twin-T,* or double-integrator oscillator, shown in Fig. 6-2* The frequency of oscillation for this circuit is given by:

$$f_o = \frac{1}{2\pi RC} \qquad \text{(Eq. 6-2)}$$

and the variable resistor $R/2$ is adjusted so that the circuit oscillates. To assure that this circuit starts immediately when the power is applied, R_2 is twice R, and R_1 is made approximately 10 times R_2.

---

* Prensky, S. D. Manual of Linear Integrated Circuits. Reston Publishing Co., Inc., Reston, 1974, p. 75.

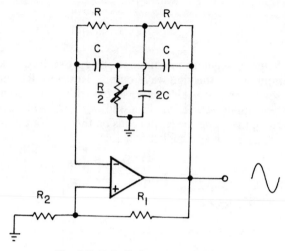

Fig. 6-2. Twin-T sine-wave oscillator.

Example

Design a 500-Hz sine-wave oscillator, using the circuit of Fig. 6-2. For example, letting C = .047 μF, then from Equation 6-2,

$$R_1 = \frac{1}{2\pi fC}$$

$$= \frac{1}{(6.28)(500 \text{ Hz})(.047 \text{ }\mu\text{F})}$$

$$= 6775 \text{ }\Omega \text{ (use 6.8 k}\Omega)$$

Capacitor 2C is then two .047-μF capacitors connected *in parallel*. Since resistor R/2 is approximately 3.4 kΩ, we can use a 5-kΩ potentiometer. To complete the design, resistor R_2 is twice R, or 13.6 kΩ, so that two 6.8-kΩ resistors connected *in series* are used. In addition, resistor R_1 is made approximately 10 times R_2, or 136 kΩ, for which a 130-kΩ resistor can be used. The completed design of the twin-T oscillator is shown in Fig. 6-3.

SINE/COSINE OSCILLATOR

If we integrate or differentiate a sine wave, as was discussed in Chapter 3, we can generate a cosine wave (i.e., a sine wave that is 90° out-of-phase). However, if we use a dual op-amp, such as a 747 or 5558 with both sections connected as integrators as shown in

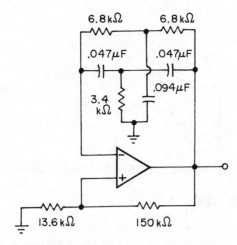

Fig. 6-3. A 500-Hz sine-wave oscillator circuit.

Fig. 6-4, we are then able to generate simultaneous *sine and cosine outputs.* Such a circuit is also called a *quadrature oscillator.*

The output frequency is given by:

$$f_o = \frac{1}{2\pi RC} \qquad \text{(Eq. 6-3)}$$

Resistor R_1 is made slightly less than R to make sure that the oscillator starts immediately when the power is applied. Usually there is a form of limiting circuitry, as shown in the partial schematic of Fig. 6-5, applied to the second op-amp A_2 to control the amplitude of the cosine output at the zener voltage, $\pm V_z$. Table 6-2 lists the common types of zener diodes and their voltages.

Table 6-2. Common Zener Diodes and Voltages

Zener Diode	V_z
1N746	3.3V
1N751	5.1
1N4734	5.6
1N4735	6.2
1N4736	6.8
1N5236	7.5
1N4738	8.2
1N757	9.0
1N4742	12.0

SQUARE AND TRIANGLE GENERATORS

The basic square-wave generator, shown in Fig. 6-6, is also called a *relaxation oscillator,* as the circuit oscillates without an external

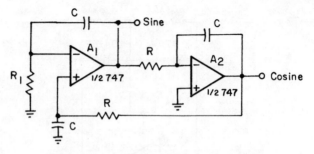

Fig. 6-4. Sine/cosine (quadrature) oscillator.

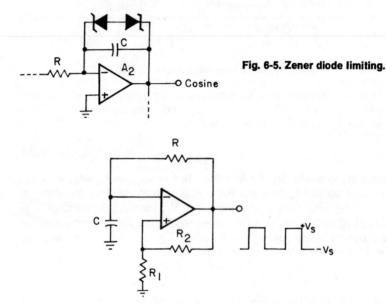

Fig. 6-5. Zener diode limiting.

Fig. 6-6. Square-wave generator.

signal. The output is fed back to both inputs, so that the output frequency is set by the charging and discharging of capacitor C through R, so that:

$$f_o \simeq \frac{1}{2RC \ln\left(\frac{2R_1}{R_2} + 1\right)} \qquad \text{(Eq. 6-4)}$$

Resistors R_1 and R_2 are chosen so that R_1 is approximately ⅓ R and R_2 is 2 to 10 times R_1.

Another oscillator that produces square waves along with simultaneous triangular waves is shown in Fig. 6-7. Op-amp A_2 is wired as

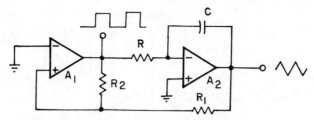

Fig. 6-7. Square/triangle wave generator.

an integrator while A_1 is essentially wired as a comparator whose reference voltage is zero.

The output amplitude of the square wave is set by the output swing of A_1, and the ratio R_1/R_2 sets the triangle's amplitude. For both waveforms, the frequency of oscillation is given by:

$$f_o = \frac{1}{4RC}\left(\frac{R_2}{R_1}\right) \qquad \text{(Eq. 6-5)}$$

With the advancement of solid-state technology, it is now common to design waveform generators with integrated circuits manufactured for this purpose. There are two chips that generate variable frequency sine, square, and triangle waveforms. One is the 8038, made by Intersil, and the other is the XR-2206, made by Exar.

Another popular chip is the 555 timer, which is capable of generating variable frequency square waves with adjustable duty cycles. In addition, triangular and sine waves are also possible with some external circuitry, as outlined in: *555 Timer Applications Sourcebook, With Experiments.**

THE STAIRCASE GENERATOR

In Fig. 6-8, we have the circuit for a linear *staircase* generator. An input square wave, having a peak-to-peak voltage V_i, charges capacitor C_1, with a charge equal to:

$$Q = CV_c$$
$$= C_1(V_i - 0.7) \qquad \text{(Eq. 6-6)}$$

When the switch across C_2 is open, capacitor C_2 is charged by each input cycle in equal voltage steps ΔV_o so that:

$$\Delta V_o = (V_i - 1.4)\frac{C_1}{C_2} \qquad \text{(Eq. 6-7)}$$

---

* Berlin, H. M. *555 Timer Applications Sourcebook, With Experiments.* Howard W. Sams & Co., Inc.

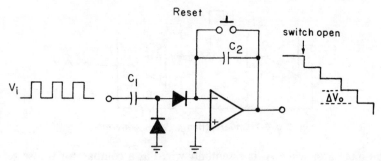

Fig. 6-8. Staircase generator.

In the limit, the maximum height of the staircase, which goes negative because of the input signal being applied to the op-amp's inverting input, is determined by the supply voltage.

AN INTRODUCTION TO THE EXPERIMENTS

The following experiments in this chapter are designed to demonstrate the use of op-amps to generate several types of periodic waveforms. Although there are presently specialized integrated circuits that are designed to do this same thing, these experiments are nevertheless intended to show the versatility of the op-amp.

The experiments that you will perform can be summarized as follows:

Experiment No.	Purpose
1	Demonstrates the design and operation of a sine/cosine (quadrature) oscillator with zener diode limiting.
2	Demonstrates the design and operation of a square-wave (relaxation) oscillator.
3	Demonstrates the design and operation of a combination square and triangle wave oscillator.

EXPERIMENT NO. 1

Purpose

The purpose of this experiment is to demonstrate the design and operation of a sine/cosine (quadrature) oscillator, using a type 741 op-amp.

Schematic Diagram of Circuit (Fig. 6-9)

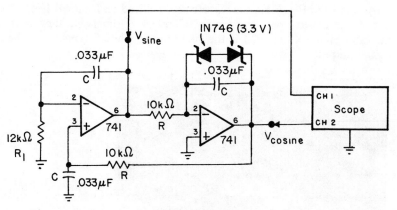

Fig. 6-9.

Design Basics

- Oscillation frequency: $f_o = \dfrac{1}{2\pi RC}$
  where $R_1 < R$
- Peak-to-peak output: V_o (p-p) $= V_z$

Step 1

Set your oscilloscope for the following settings:
- Channels 1 & 2: 1 volt/division
- Time base: 0.5 msec/division
- AC coupling

Step 2

Wire the circuit shown in the schematic diagram. Be sure to check the polarity of the two zener diodes. Apply power to the breadboard and position the Channel 1 trace above the Channel 2 trace. For each trace, determine the oscillation frequency. Are they both the same?

The frequency of the trace of Channel 1 is exactly the same as that of Channel 2. The calculated frequency, using the components given in the schematic diagram, is 482 Hz. When this experiment was performed, a frequency of 370 Hz was measured.

Step 3

How does the waveform of Channel 1 compare with the waveform of Channel 2?

139

As shown in Fig. 6-10, the waveform displayed on Channel 1 is 90° out-of-phase with the waveform of Channel 2; otherwise they have the same shape. Channel 1 is the sine output, while Channel 2 is the cosine output.

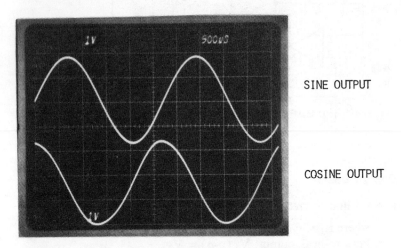

SINE OUTPUT

COSINE OUTPUT

Fig. 6-10. Output waveforms for sine/cosine oscillator in Fig. 6-9 when R_1 is 12 kΩ.

Step 4

Measure the peak-to-peak voltage of the cosine-wave (Channel 2). Since we are using 3.3 volt diodes, is the peak-to-peak voltage approximately equal to the diode's rating?

Step 5

Disconnect the power from the breadboard. Now change the resistor R_1 to 8.2 kΩ. Apply power to the breadboard. What do you notice about the *cosine* waveform?

In our case, the cosine wave is distorted, as shown in Fig. 6-11, compared with the tracing shown in Step 3.

Step 6

Measure the output frequency. It is different from the value you measured in Step 2?

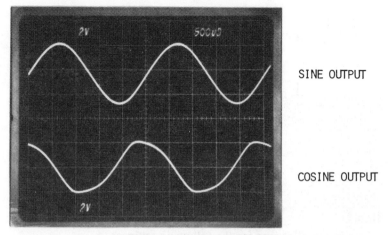

SINE OUTPUT

COSINE OUTPUT

Fig. 6-11. Output waveforms for sine/cosine oscillator in Fig. 6-9 when R_1 is 8.2 kΩ.

In our case, the output frequency is now 392 Hz, or 22 Hz higher than in Step 2. No explanation can be offered for this difference, since R_1 is not a factor in the "frequency" equation. You may have also noticed that we originally used a value for R_1 that was greater than the 10-kΩ frequency-determining resistor, thereby contradicting the design equation. With this higher value, we obtained a better looking cosine wave. Again, no explanation can be offered. Therefore, I suggest that you experimentally determine the best value for R_1 to give an undistorted cosine wave at the output of the 2nd op-amp, using the design criterion in the *Design Basics* section as a starting point.

EXPERIMENT NO. 2

Purpose

The purpose of this experiment is to demonstrate the design and operation of a square-wave (relaxation) oscillator, using a type 741 op-amp.

Schematic Diagram of Circuit (Fig. 6-12)

Design Basics

- Oscillation frequency: $f_o \simeq \dfrac{1}{2RC \ln\left(\dfrac{2R_1}{R_2} + 1\right)}$

  where, $R_1 \simeq \dfrac{R}{3}$

  $R_2 = 2$ to $10\,R_1$

141

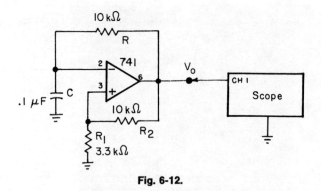

Fig. 6-12.

Step 1

Set your oscilloscope for the following settings:
- Channel 1: 5 volts/division (for ±15-volt supply)
              2 volts/division (for ±6-volt supply)
- Time base: 0.2 msec/division
- AC coupling

Step 2

Wire the circuit shown in the schematic diagram (Fig. 6-12). Apply power to the breadboard. You should see a square wave. What is the frequency of oscillation?

Within 10%, you should have measured a frequency near 987 Hz. When this experiment was performed, a frequency of 909 Hz was measured.

Step 3

Disconnect the power from the breadboard. Change the capacitor C to 0.0047 μF and the scope's time base to 50 μsec/division. Then apply power to the breadboard. What do you notice about the output waveform?

The output waveform should resemble a triangle wave, but with flat positive and negative peaks, as shown in Fig. 6-13. Because of the slew rate of the 741 op-amp (typically 0.5 V/μsec), the output square wave then becomes distorted. With this 0.0047-μF capacitor, the output frequency should be about 21 kHz, beyond the acceptable limit for the 741 op-amp.

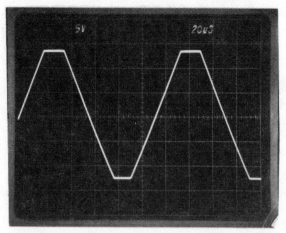

Fig. 6-13. Output waveform for square-wave oscillator in Fig. 6-12.

Step 4

As accurately as you can, measure the time it takes for the output signal to switch from its maximum negative voltage to its maximum positive voltage. Record both the measured time period and the peak-to-peak output swing below:

$$\text{time} = \underline{\hspace{1cm}} \ \mu\text{sec}$$

$$\text{voltage} = \underline{\hspace{1cm}} \ \text{volts}$$

Step 5

From the two measurements of Step 4, determine the slew rate for the type 741 op-amp from the expression:

$$\text{slew rate} = \frac{\text{voltage}}{\text{time}}$$

$$= \underline{\hspace{1cm}} \ \text{V}/\mu\text{sec}$$

From this calculation, how does your value compare with the manufacturer's published value?

When I performed this experiment, I measured a time of 42 μsec for the output swing of 26 volts, giving a slew rate of 0.62 V/μsec, compared with a typical value of 0.5 V/μsec.

Step 6 (Optional)

If you have a high slew rate op-amp, such as the LM318 (70 V/μsec), disconnect the power from the breadboard and use this op-amp in place of the 741 op-amp (the pin connections are the same!). Apply power to the breadboard. How does the waveform's shape compare with the one you obtained in Step 3?

The waveform should appear as a normal square wave; for this type of circuit, the op-amp used should have a high slew rate for reliable results.

EXPERIMENT NO. 3

Purpose

The purpose of this experiment is to demonstrate the design and operation of a combination square and triangle wave generator, using two type 741 op-amps.

Schematic Diagram of Circuit (Fig. 6-14)

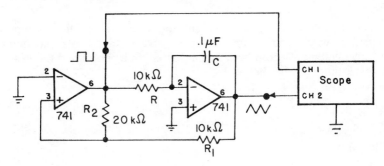

Fig. 6-14.

Design Basics

- Frequency of oscillation: $f_o = \dfrac{1}{4RC}\left(\dfrac{R_2}{R_1}\right)$

- Based on the peak-to-peak square-wave output $(V_S)_{p\text{-}p}$, the peak-to-peak triangle output is:

$$(V_T)_{p\text{-}p} = V_S \frac{R_1}{R_2}$$

Step 1

Set your oscilloscope for the following settings:
- Channel 1: 5 volts/division (for a ±15-volt supply)
      2 volts/division (for a ±6-volt supply)
- Channel 2: 2 volts/division (±15-volt supply)
      1 volt/division (±6-volt supply)

- Time base: 1 msec/division
- AC coupling

Step 2

Wire the circuit shown in the schematic diagram (Fig. 6-14). Apply power to the breadboard. You should now see a square wave and a triangle wave on the scope screen. Measure the peak-to-peak voltage of the square wave, and record it below:

$$(V_S)_{p-p} = \underline{\hspace{1cm}} \text{ volts}$$

Naturally, this peak-to-peak value will depend on the supply used. Using a ±15-volt supply, I measured a voltage of 26 volts; using a ±6-volt supply, this value was about 10 volts.

Step 3

Now measure the peak-to-peak voltage of the triangle wave, recording your result below. How does it compare with the equation given in the *Design Basics* section?

$$(V_T)_{p-p} = \underline{\hspace{1cm}} \text{ volts}$$

When this experiment was performed using a ±15-volt supply, the peak-to-peak triangle voltage was 13.4 volts, as compared with a calculated value of 13.0 volts.

Step 4

Now measure the output frequency. How does it compare with the design equation?

Within 10%, it should be about 500 Hz.

Step 5

Verify the design equations by changing R_1 to 4.7 kΩ and then to 15 kΩ, repeating Steps 1 through 4.

CHAPTER 7

Active Filters

INTRODUCTION

In this chapter, several of the major filter configurations using op-amps will be presented. This chapter, however, will be limited to 2nd order low-pass and high-pass Butterworth-type responses. A more comprehensive presentation on the practical design of active filters can be found in *Design of Active Filters, With Experiments* or *Active-Filter Cookbook,* both published by Howard W. Sams & Co., Inc.

OBJECTIVES

At the completion of this chapter, you will be able to do the following:

* Design and predict the performance of the following filters:
  A 2nd order "equal component" low-pass active filter.
  A 2nd order "equal component" high-pass active filter.
  A multiple feedback bandpass filter.
  A state-variable filter with simultaneous low-pass, high-pass, and
      bandpass output responses.
  A state-variable notch filter.

WHAT IS AN ACTIVE FILTER?

"A filter is a device or substance that passes electric currents at certain frequencies or frequency ranges while preventing the passage of others" (Webster). Specifically, an *active* filter is a network composed of resistors and capacitors built around an op-amp. An active filter offers the advantages of:

- *No insertion loss:* Since the op-amp is capable of providing gain, the input signal will not be immediately attenuated while the filter passes those frequencies of interest.
- *Cost:* Active filters, on the average, will cost far less than passive-type filters. This is because inductors are expensive and are not always readily available.
- *Tuning:* Active filters are easily tuned, or adjusted over a wide frequency range without altering the desired response.
- *Isolation:* As a result of using an op-amp, active filters will have a high input impedance and a low output impedance, virtually guaranteeing almost no interaction between the filter and its source or load.

On the other hand, there are several disadvantages, or limitations with using active filters:

- *Frequency response:* You are at the mercy of the type of op-amp used in your design (see Chapter 1).
- *Power supply:* Unlike passive filters, active filters require some form of power supply for the op-amp.

2nd ORDER VCVS FILTERS

The simplest 2nd order *low-pass* active filter is the voltage-controlled-voltage-source (VCVS) circuit of Fig. 7-1, which is also referred to as the *Sallen and Key* filter. For this circuit, the cutoff frequency is given by:

$$f_c = \frac{1}{2\pi(R_1R_2C_1C_2)^{1/2}} \qquad \text{(Eq. 7-1)}$$

The big decision now facing us is how do we pick the values for the two resistors and capacitors that make up Equation 7-1? The

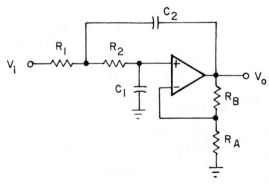

Fig. 7-1. A 2nd order VCVS filter.

easiest approach is to make R_1 and R_2 equal, and C_1 and C_2 equal, so that Equation 7-1 now becomes:

$$f_c = \frac{1}{2\pi RC} \qquad \text{(Eq. 7-2)}$$

This is called an "equal-component" VCVS low-pass filter. The pass-band gain is fixed at 1.586 (+4 dB) for a 2nd order Butterworth response, and *this is the only gain that will permit this circuit to work.* The cutoff frequency will be at the point where the filter's response is 3 dB less than the passband gain of +4 dB, or +1 dB. Beyond the cutoff frequency the response *decreases* by a rate of 12 dB/octave, or 20 dB/decade.

Since this filter uses the op-amp in the noninverting mode, the feedback resistor R_B must be 0.586 times the value of the input resistor R_A for a voltage gain of 1.586. Using ±5% resistors, a good choice for these resistors is 27 kΩ and 47 kΩ, respectively, as shown in Fig. 7-2.

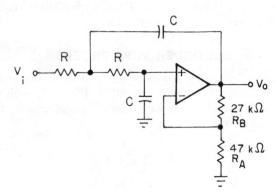

Fig. 7-2. Equal-component VCVS low-pass filter.

Example

Design a 2nd order Butterworth VCVS low-pass active filter with a cutoff frequency of 700 Hz.

Since it is easier to start by selecting capacitor values, assume C is, for example, 0.0033 μF. Then by Equation 7-2, R is found:

$$R = \frac{1}{2\pi f_c C}$$

$$= \frac{1}{(2\pi)(700 \text{ Hz})(.0033 \text{ μF})}$$

$$= 68,898 \text{ Ω (use 68 kΩ resistor)}$$

so that the final circuit is shown in Fig. 7-3.

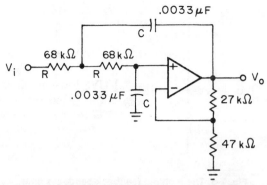

Fig. 7-3. Equal-component VCVS low-pass filter showing values for R and C.

Of course, we could have chosen any other standard capacitor value, such as 0.01 μF, in which case, R is 22,736 Ω (use a 22 kΩ resistor).

By simply interchanging the position of the frequency-determining components of the circuit of Fig. 7-2, we then obtain a 2nd order "equal-component" VCVS *high-pass* active filter, as shown in Fig. 7-4. Like the low-pass version, the cutoff frequency is the same as that given by Equation 7-2, and the passband gain is also 1.586, or +4 dB.

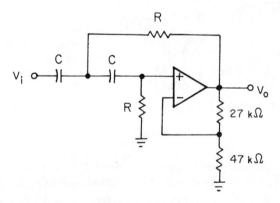

Fig. 7-4. Equal-component Butterworth VCVS high-pass filter.

THE MULTIPLE FEEDBACK BANDPASS FILTER

As shown in Fig. 7-5, the basic multiple feedback bandpass filter is applicable for Qs approximately less than 10. There is an additional feedback path, hence the term "multiple feedback." In addition, the op-amp is connected in the *inverting* mode.

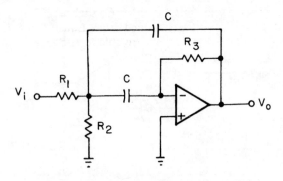

Fig. 7-5. The multiple feedback bandpass filter.

For this circuit, the *center frequency* is determined from the relation:

$$f_o = \frac{1}{2\pi C}\left[\frac{1}{R_3}\frac{R_1 + R_2}{R_1 R_2}\right]^{1/2} \qquad \text{(Eq. 7-3)}$$

However, the determination of the three resistors is greatly simplified by the following equations:

$$R_1 = \frac{Q}{2\pi f_o G_o C} \qquad \text{(Eq. 7-4)}$$

$$R_2 = \frac{Q}{2\pi f_o C(2Q^2 - G_o)} \qquad \text{(Eq. 7-5)}$$

and,

$$R_3 = \frac{Q}{\pi f_o C} \qquad \text{(Eq. 7-6)}$$

However, Equations 7-4 and 7-6 can be combined to give:

$$G_o = \frac{R_3}{2R_1} \qquad \text{(Eq. 7-7)}$$

On the other hand, because of the denominator of Equation 7-5, we are restricted by the interrelation between G_o and Q so that:

$$Q > (G_o/2)^{1/2} \qquad \text{(Eq. 7-8)}$$

Normally we select a convenient value for C and then systematically calculate the required values for the three resistors based upon the requirements for Q, G_o, and f_o.

Example

Design a 750-Hz bandpass filter using the circuit of Fig. 7-5 with a center frequency gain of 1.32, and a Q of 4.2.

First we select a standard value for C, such as .01 μF. Then the three resistors are in turn determined from Equations 7-4, 7-5, and 7-7. Since Equation 7-8 holds, we are able to continue, so that:

$$R_1 = \frac{Q}{2\pi f_o G_o C}$$
$$= \frac{(4.2)}{(2\pi)(750 \text{ Hz})(1.32)(.01 \, \mu\text{F})}$$
$$= 67.6 \text{ k}\Omega \text{ (use 68 k}\Omega\text{)}$$

$$R_3 = 2R_1 G_o$$
$$= (2)(67.6 \text{ k}\Omega)(1.32)$$
$$= 178 \text{ k}\Omega \text{ (use 180 k}\Omega\text{)}$$

$$R_2 = \frac{Q}{2\pi f_o C(2Q^2 - G_o)}$$
$$= \frac{(4.2)}{(2\pi)(750 \text{ Hz})(.01 \, \mu\text{F})[(2)(4.2)^2 - 1.32]}$$
$$= 2.6 \text{ k}\Omega \text{ (use 2.7 k}\Omega\text{)}$$

The final circuit is shown below in Fig. 7-6.

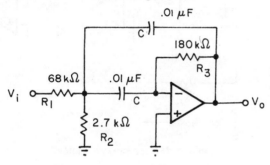

Fig. 7-6. A 750-Hz bandpass filter.

One nice feature about this circuit is that the center frequency f_o can be easily shifted to a new value f_o' simply by changing resistor R_2 to a new value R_2'. In this way, the center frequency gain and the bandwidth both remain constant. The new value for R_2 is computed from the formula,

$$R_2' = R_2\left(\frac{f_o}{f_o'}\right)^2 \qquad \text{(Eq. 7-9)}$$

Example

Using the previous example, change the center frequency of the filter from 750 Hz to 600 Hz.

To change the center frequency from 750 Hz to 600 Hz, a new value for R_2 is calculated from Equation 7-9 so that:

$$R_2' = 2.7 \text{ k}\Omega\left(\frac{750 \text{ Hz}}{600 \text{ Hz}}\right)^2$$

$$= 4.2 \text{ k}\Omega \text{ (use 4.3 k}\Omega\text{)}$$

THE STATE VARIABLE FILTER

By properly connecting three op-amps we are able to simultaneously provide 2nd order low- and high-pass, and bandpass filter outputs. Such a filter is shown in Fig. 7-7, and is called a *state variable filter,* or a *universal active* filter. As seen from the circuit, it is basi-

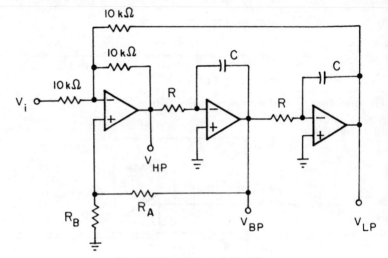

Fig. 7-7. The state-variable filter.

cally composed of a difference amplifier (A_1) and two identical integrators (A_2 and A_3). For such a filter, the center and cutoff frequencies are the same, as given by the formula:

$$f_o \text{ or } f_c = \frac{1}{2\pi RC} \qquad \text{(Eq. 7-10)}$$

The filter's Q is set solely by resistors R_A and R_B,

$$R_A = (3Q - 1)R_B \qquad \text{(Eq. 7-11)}$$

For the low-pass and high-pass outputs, the passband gain is *unity.* For the bandpass output, the center frequency gain will be equal to the value of Q.

As this type of circuit gives a 2nd order response for both the low-pass and high-pass outputs, *it will not be possible to obtain optimum performance with all three outputs simultaneously.* For either a low-pass or high-pass *Butterworth* response, Q must be equal to 0.707. Consequently the bandpass response suffers terribly! Even for a 2nd order 3-dB Chebyshev filter, Q is 1.3, which is not much better. We should then design either for a 2nd order Butterworth low-pass/high-pass response (Q = 0.707), or a high-Q bandpass response.

Example

Design a state variable filter with a center frequency of 60 Hz and a Q of approximately 50, using the circuit of Fig. 7-7.

Using Equation 7-10, and choosing C equal to 0.22 μF, for example, then:

$$R = \frac{1}{2\pi f_o C}$$
$$= \frac{1}{(2\pi)(60\ \text{Hz})(0.22\ \mu\text{F})}$$
$$= 12{,}057\ \Omega\ (\text{use } 12\ \text{k}\Omega)$$

Since Q = 50, resistors R_A and R_B are found from Equation 7-11,

$$R_A = (3Q - 1)R_B$$
$$= [(3)(50) - 1]R_B$$
$$= 149\ R_B$$

so that R_A is 149 times larger than R_B. If R_B is equal to 1 kΩ, then R_A is 149 kΩ, for which we can use a 150-kΩ resistor for all practical purposes, resulting in the completed design shown in Fig. 7-8.

THE NOTCH FILTER

One very nice feature of the state variable filter of Fig. 7-7 is that we can *simultaneously add the low-pass and high-pass outputs equally,* thus creating a notch, or band reject filter. Such a filter is extremely useful for minimizing the presence of 60-Hz "hum" on audio signals. What is needed now is a 2-input summing amplifier with equal gains, as shown in Fig. 7-9.

Example

Using the previous example (Fig. 7-8), design a 60-Hz notch filter with a Q of 50.

From the previous example, everything from the basic 60-Hz state variable filter of Fig. 7-8 remains the same. All that is needed now is the addition of the circuit of Fig. 7-9.

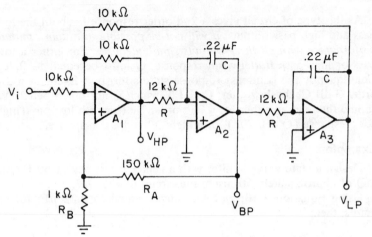

Fig. 7-8. State-variable filter with center frequency of 60 Hz and Q of approximately 50.

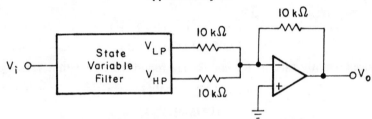

Fig. 7-9. State-variable notch filter.

AN INTRODUCTION TO THE EXPERIMENTS

The following experiments are designed to acquaint you with the design, operation, and testing of active filters. If you wish to learn more about other filter configurations, refer to *Design of Active Filters, With Experiments,* published by Howard W. Sams & Co., Inc., which contains over 25 experiments.

The experiments that you will perform can be summarized as follows:

Experiment No.	Purpose
1	Demonstrates the design and operation of an "equal-component" VCVS Butterworth low-pass active filter.
2	Demonstrates the design of a multiple feedback bandpass filter with frequency shifting.
3	Demonstrates the design and operation of a unity-gain state-variable filter.

EXPERIMENT NO. 1

Purpose

The purpose of this experiment is to demonstrate the design and operation of an "equal-component" VCVS Butterworth low-pass filter using a 741 op-amp.

Schematic Diagram of Circuit (Fig. 7-10)

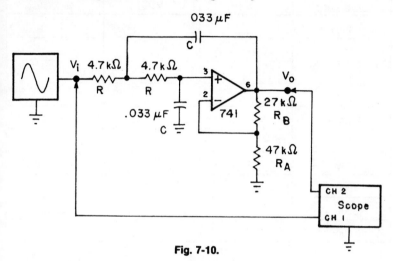

Fig. 7-10.

Design Basics

- Cutoff frequency: $f_c = \dfrac{1}{2\pi RC}$
- Gain: fixed at 1.586 by R_A and R_B for a 2nd order Butterworth response.
- Amplitude response: $20 \log_{10} \left(\dfrac{1.586}{[1 + (f/f_c)^4]^{1/2}} \right)$

Step 1

Set your oscilloscope for the following settings:
- Channel 1: 0.2 volt/division
- Channel 2: 0.5 volt/division
- Time base: 1 msec/division
- AC coupling

Step 2

Wire the circuit shown in the schematic diagram (Fig. 7-10). Apply power to the breadboard and adjust the input voltage at 1 volt peak-to-peak and a frequency of 100 Hz.

Step 3

Vary the generator's frequency and complete the following table, plotting your results on the blank graph provided in Fig. 7-11 for this purpose. Keep the input voltage constant over this frequency range!

Frequency	V_o	V_o/V_i	dB Gain
100 Hz			
200			
400			
600			
800			
1000			
2000			
4000			
8000			
10,000			

Step 4

From your results in Step 3, what is your measured passband gain?

You should have measured a passband gain (at 100 Hz) of approximately 1.58 (+4 dB). This is the only value that will permit this 2nd order low-pass filter to exhibit a Butterworth, or "maximally flat" response.

Step 5

From your graph, determine the filter's cutoff frequency. How does it compare with the design equation?

The cutoff frequency is the point at which the amplitude response is 3 dB less than the passband gain (Step 4), or +1 dB. Within 10% this value should be approximately equal to the design frequency of 1026 Hz.

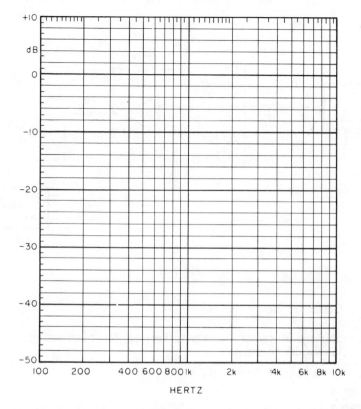

Fig. 7-11.

EXPERIMENT NO. 2

Purpose

The purpose of this experiment is to demonstrate the design and operation of a simple bandpass filter using a type 741 op-amp.

Schematic Diagram of Circuit (Fig. 7-12)

Design Basics

- Center frequency: $f_o = \dfrac{1}{2\pi C}\left[\dfrac{1}{R_3}\left(\dfrac{1}{R_1} + \dfrac{1}{R_2}\right)\right]^{1/2}$

where, $R_1 = \dfrac{Q}{2\pi f_o GC}$

$R_2 = \dfrac{Q}{2\pi f_o C(2Q^2 - G)}$

$R_3 = \dfrac{2Q}{2\pi f_o C}$

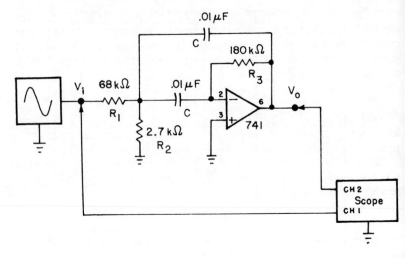

Fig. 7-12.

- Center frequency gain: $G = \dfrac{R_3}{2R_1}$

- Shifting of center frequency, keeping the passband gain and bandwidth constant:

$$R_2' = R_2\left(\frac{f_o}{f_o'}\right)^2$$

Step 1

Set your oscilloscope for the following settings:

- Channels 1 & 2: 0.2 volt/division
- Time base: 0.2 msec/division
- AC coupling

Step 2

Wire the circuit shown in the schematic diagram (Fig. 7-12). Apply power to the breadboard and adjust the input voltage at 1.4 volts (i.e., 7 vertical divisions). Make this setting as accurate as you can.

Step 3

Now vary the generator frequency so that the output voltage, as displayed on Channel 2 of your oscilloscope, reaches its *maximum amplitude*. Measure this output voltage and determine the voltage gain at this point. It may be necessary to reduce the sensitivity of Channel 2 to 0.5 volt/division. How does your measured gain compare with the expected value?

Within 10%, the gain should be 1.32 (I measured 1.28). Next count the number of horizontal divisions occupied by one complete cycle, without touching the generator frequency control, in order to determine the filter's center frequency. How does it compare with the theoretical value?

Your measured frequency should be near 737 Hz. If you don't know how to determine the frequency from the duration of one complete cycle as seen on the scope screen, consider my results:

When I performed this experiment, one complete cycle occupied 6.6 horizontal divisions. Since the time base is set at 0.2 msec/divisions, then the frequency is found from:

$$f = \frac{1}{(6.6 \text{ divisions})(0.2 \text{ msec/division})}$$

$$= \frac{1}{1.32 \text{ msec}}$$

$$= 758 \text{ Hz}$$

Step 4

Now determine the upper and lower 3-dB frequencies by measuring the two frequencies at which the amplitude response decreases by a factor of 0.707 of the center frequency gain (-3 dB). To do this, multiply your measured center frequency gain by 0.707, and then multiply this intermediate value by 1.4 (the input voltage) to obtain the output voltage at which the response is 3 dB less than the center frequency. After you have determined this value, adjust the input frequency above and below the center frequency until the peak-to-peak voltage reaches the particular value that you have just calculated. Record your results below:

$$f_L = \underline{\hspace{2cm}} \text{ Hz}$$

$$f_H = \underline{\hspace{2cm}} \text{ Hz}$$

Step 5

Subtract the lower value from the higher value to determine the 3-dB bandwidth. Then divide this value into the value you determined as the center frequency (Step 3), which is the Q, or *quality factor,*

$$Q = \frac{f_o}{f_H - f_L}$$

$$= \underline{\hspace{2cm}}$$

Within 10%, you should have determined a Q of 4.17. If not, repeat Steps 3 and 4, carefully measuring the voltages and frequencies. In our case, the measured 3-dB frequencies were 672 and 860 Hz, so that the bandwidth was 188 Hz, and the filter's Q was 758/188, or 4.03.

Step 6

Disconnect the power from the breadboard and replace the 2.7-kΩ resistor (R_2) with a 1.5-kΩ resistor. Reconnect the power to the breadboard and adjust the input voltage now at 1.0 volt peak-to-peak. Repeat Steps 3, 4, and 5 to determine the filter's voltage gain, center frequency, bandwidth, and Q, completing the following table:

gain = _____

f_H = _____ Hz

f_L = _____ Hz

bandwidth = _____ Hz

center frequency = _____ Hz

Q = _____

Step 7

By changing the value of R_2, does the new center frequency that you have just determined compare favorably with the equation given in the *Design Basics* section?

From the equation, the new center frequency should be approximately:

$$f_o' = (758 \text{ Hz}) \left[\frac{2.7 \text{ k}\Omega}{1.5 \text{ k}\Omega} \right]^2$$

$$= 1017 \text{ Hz}$$

which is based on the center frequency that I determined in Step 3.

If you have done everything correctly and carefully, you should find that the filter's bandwidth, as found in Step 6, should approximately be the same as the value you determined in Step 5 (the original circuit), even though the center frequency has been changed. Consequently, the changing of R_2 changes only the center frequency. The filter's gain and bandwidth essentially remain constant. If you are still not convinced, try another resistance for R_2 and repeat Step 6.

EXPERIMENT NO. 3

Purpose

The purpose of this experiment is to demonstrate the design, operation, and characteristics of a unity gain state-variable filter, using three type 741 op-amps.

Schematic Diagram of Circuit (Fig. 7-13)

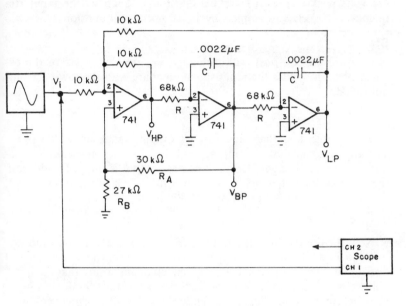

Fig. 7-13.

Design Basics

- Cutoff frequency: $f_c = 1/2\pi RC$
- $R_A = (3Q - 1)R_B$
- Passband gains:
    - Low-pass $= -1$ ($180°$ out of phase)
    - High-pass $= -1$ ($180°$ out of phase)
    - Bandpass $= Q$ (in phase)

Step 1

Set your oscilloscope for the following settings:

- Channels 1 & 2: 0.2 volt/division
- Time base: 1.0 msec/division
- AC coupling

Step 2

Wire the circuit shown in the schematic diagram (Fig. 7-13). If you don't have a 30-kΩ resistor, use two 15-kΩ resistors *connected in series* instead.

Step 3

Apply power to the breadboard and first adjust the input signal at 1.4 volts peak-to-peak (7 vertical divisions). Then set the input frequency at 100 Hz (1 complete cycle/10 horizontal divisions).

Step 4

Measure the *low-pass* output voltage V_{LP} on Channel 2 of the oscilloscope. How does the input signal compare with the low-pass output signal?

You should find that the low-pass output signal amplitude is essentially the same as the input, or 1.4 volts peak-to-peak. Consequently, the voltage gain for the low-pass output is unity. In addition, the low-pass output signal is inverted with respect to the input signal, so that the two are out of phase by 180° in the filter's passband.

Step 5

Now set the oscilloscope time base at 0.1 msec/division, and adjust the input frequency so that 1 complete cycle occupies the 10 horizontal divisions (10 kHz). Also, check to make sure that the input voltage is still 1.4 volts peak-to-peak. Then measure the filter's high-pass output V_{HP} on Channel 2. How does the input signal compare with the low-pass output?

As with the low-pass output, the peak-to-peak high-pass output voltage should be approximately 1.4 volts, so that the voltage gain is also unity. In addition, the high-pass output signal is inverted with respect to the input, indicating a 180° phase shift in the high-pass filter's passband.

Step 6

Decrease the input frequency until the voltage of the high-pass output reaches 1.0 volt peak-to-peak, which is 0.707 times the input voltage, or a decrease of 3 dB. Measure this frequency as accurately as you can and record it below:

$$f_c \text{ (HP)} = \underline{\hspace{2cm}} \text{ Hz}$$

Step 7

Without disturbing the setting of the frequency generator, transfer the probe connected to the high-pass output to the low-pass output of the filter, and measure the peak-to-peak voltage. Is it 1.0 volt? If not, vary the generator frequency slightly so that the output voltage is now 1.0 volt peak-to-peak. Measure this frequency as accurately as you can and record it below:

$$f_c \text{ (LP)} = \text{_____ Hz}$$

If the frequency-determining components of both integrators are fairly well matched, the frequencies measure in Steps 6 and 7 should almost be the same, to within several Hz. How do these two frequencies compare with the expected cutoff frequency?

Within 10%, both frequencies should be near the theoretical value of 1064 Hz.

Step 8

Now transfer the Channel 2 probe of the scope to the filter's band-pass output V_{BP}. Very carefully vary the frequency up and down, stopping at the point at which the output voltage reaches a maximum value. Now measure this frequency as carefully as you can and record it below:

$$f_o \text{ (BP)} = \text{_____ Hz}$$

If the two measured frequencies in Steps 6 and 7 are different, indicating that the two integrators are mismatched to some degree, we must then determine the expected center frequency based on the measurements of Steps 6 and 7. To do this, we take the *geometric average,* so that:

$$f_o = (f_{LP}f_{HP})^{1/2}$$

When I performed this experiment, I measured frequencies of 971 Hz and 968 Hz for Steps 6 and 7, respectively. Therefore, the geometric average for f_o is 969.5 Hz, while the measured value for f_o in this step was 970 Hz. If you found that the frequencies in Steps 6 and 7 were equal, how does it compare with the measured value in this step?

Within a few Hz, they should be the same.

Step 9

Without disturbing anything, measure the filter's bandpass output voltage V_{BP}. What is the voltage gain at the filter's center frequency? Is it what you would expect?

You should have measured a peak-to-peak voltage of approximately 1.0 volt, so that the voltage gain will be 0.70 at this center frequency which is numerically equal to the filter's Q, which is supposed to be .707. Consequently, we should conclude that this state-variable filter is primarily designed to be used either as a low-pass or a high-pass filter with Butterworth responses. From the equation in the *Design Basics* section relating resistors R_A and R_B, we find that Q should be equal to 0.704 with the component values shown. In addition, you should observe that the bandpass output signal at the center frequency is in phase with the input.

Step 10

Disconnect the power from the breadboard. Now replace the 30-kΩ resistor (R_A) with a 270-kΩ resistor and apply the power to the breadboard. Repeat Steps 1 through 9 of this experiment. How do your results compare with the design equations?

You should find no change in the frequencies of Steps 6, 7, and 8, or the passband gain for either the low-pass or high-pass outputs. However, the filter's Q has increased to approximately 3.7 and is also the bandpass filter's gain at the center frequency.

CHAPTER 8

Single Supply Operation

INTRODUCTION

Although primarily designed to be powered from a bipolar or dual power supply, a number of op-amp circuits can nevertheless be effectively operated from a single supply voltage. When used in this manner, the circuit's quiescent dc output voltage is usually set at one-half the supply voltage. Consequently, op-amp circuits using this approach always require ac coupling capacitors for the input and output signals.

OBJECTIVES

At the completion of this chapter, you will be able to do the following:

- Using single supply biasing, design and predict the performance of the following circuits:

    an inverting amplifier
    a summing amplifier
    a difference amplifier

SINGLE SUPPLY BIASING

The previous chapters discuss op-amp circuits that are primarily designed to be powered by a bipolar supply. This is particularly evident with the linear amplifier circuits presented in Chapter 2, where the input signal voltage varies both above and below ground potential.

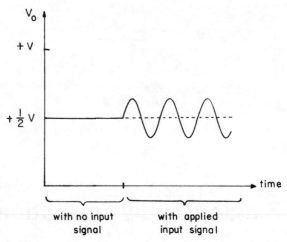

Fig. 8-1. The dc output voltage is one-half the supply voltage with no
signal applied.

If the op-amp is to properly operate from a single supply voltage, *the circuit must be able to produce negative-going as well as positive-going signals*. The best approach is then to fix the circuit's dc output voltage at one-half the supply voltage with no input signal applied (Fig. 8-1). This is called a class-A amplifier. When an input signal, such as a sine wave, is applied to the circuit, the output signal will then vary about the dc, or *quiescent* (resting) voltage. At this point the output signal is a superposition of a quiescent *dc component* and an *ac component* (the actual amplified input signal). The dc component can be thought of as an output offset voltage. For proper amplifier operation, we would like to be able to remove the dc component.

In the following sections of the chapter, we will learn how to properly bias several types of linear amplifier circuits to permit operation from a single supply voltage.

THE INVERTING AMPLIFIER

The basic inverting amplifier circuit using a single supply is shown in Fig. 8-2. Comparing this circuit with the standard bipolar supply circuit of Fig. 2-2, we notice several differences. First, the power supply connections to the op-amp are included in the schematic diagram to indicate that a single supply voltage $(+V)$ is used; *the $-V$ op-amp connection now goes directly to ground*. Secondly, instead of grounding the noninverting input, the voltage divider R3-R4 provides a voltage to the noninverting input to set the op-amp's output at $+V/2$ with no input signal. Finally, capacitors C_1 and C_2 are

166

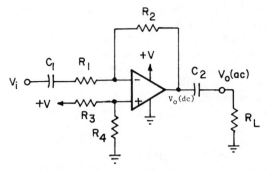

Fig. 8-2. Inverting amplifier using a single supply voltage.

used to couple the input and output signals, respectively, to and from the circuit, removing the dc offset.

As with the circuit of Fig. 2-2, the voltage gain of this single supply ac inverting amplifier is given by:

$$\text{voltage gain} = \frac{V_o}{V_i}$$

$$= -\frac{R_2}{R_1} \qquad \text{(Eq. 8-1)}$$

The resistors R_3 and R_4 are, in general, made equal to each other so that the quiescent dc output level will be set at $+V/2$. However, the quiescent dc output voltage is given by:

$$(V_o)_{dc} = \frac{R_4}{R_3 + R_4}(+V) \qquad \text{(Eq. 8-2)}$$

The values of the coupling capacitors C_1 and C_2 are determined by the desired low-frequency response, and either the input impedance of the circuit (for C_1) or the load (for C_2). The input coupling capacitor is found from,

$$C_1 = \frac{1}{2\pi f_c R_1} \qquad \text{(Eq. 8-3)}$$

and C_2 is determined in a like manner, substituting R_L for R_1 so that:

$$C_2 = \frac{1}{2\pi f_c R_L} \qquad \text{(Eq. 8-4)}$$

In both equations, f_c is the lowest expected input frequency.

Example

Design an ac inverting amplifier using the circuit of Fig. 8-2 with a +15-volt supply. The circuit has a voltage gain of 10, an input

167

impedance of 10 kΩ, a low-frequency response of approximately 30 Hz, and drives a 1-kΩ load.

For an input impedance of 10 kΩ, then $R_1 = 10$ kΩ. Consequently, R_2 must be equal to 100 kΩ for a voltage gain of 10. In general, resistors R_3 and R_4 can be any value we like, as long as they are *equal*. An often-used guideline is to make these two resistors approximately twice that of R_2. In this case, R_3 and R_4 are both 200 kΩ.

Using Equation 8-3, the input coupling capacitor is calculated:

$$C_1 = \frac{1}{2\pi f_c R_1}$$

$$= \frac{1}{(6.28)(30 \text{ Hz})(10 \text{ k}\Omega)}$$

$$= 0.5 \ \mu\text{F}$$

In a similar manner, using Equation 8-4, the output capacitor C_2 is found to be 5 μF. The completed design is shown in Fig. 8-3.

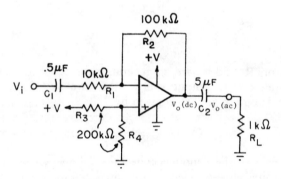

Fig. 8-3. An inverting amplifier with a gain of 10 and an input impedance of 10 kΩ that uses a single supply voltage.

Since the quiescent dc output voltage at $V_{o(dc)}$ is usually made equal to one-half the supply voltage, for example +15 volts, the maximum peak-to-peak output voltage that we can expect without distortion is also 15 volts. Consequently, the maximum peak-to-peak input voltage that can be applied is:

$$(V_i)_{\text{maximum, peak-to-peak}} = \frac{+V}{\text{voltage gain}} \qquad \text{(Eq. 8-5)}$$

For the previous example, the maximum peak-to-peak input voltage that can be applied without distortion is 1.5 volts.

THE SUMMING AMPLIFIER

As shown below in Fig. 8-4, we can place additional resistors at the op-amp's inverting input so that the resulting circuit will add these input signals. For this circuit, the output voltage is the same as Equation 2-14, or:

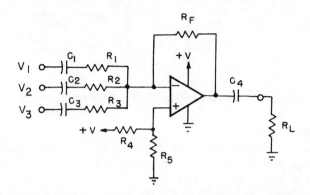

Fig. 8-4. Summing amplifier using a single supply voltage.

$$V_o = - R_F\left(\frac{V_1}{R_1} + \frac{V_2}{R_2} + \frac{V_3}{R_3}\right) \qquad \text{(Eq. 8-6)}$$

As before, resistors R_4 and R_5 are made equal to bias the quiescent dc output voltage at one-half the supply voltage. The input and output coupling capacitors are determined in the same manner as for the inverting amplifier circuit of Fig. 8-2.

THE DIFFERENCE AMPLIFIER

The difference amplifier that is powered by a single voltage is shown on Fig. 8-5 on the next page. Like the almost similar bipolar supply circuit, the output voltage in terms of the circuit's resistances is:

$$V_o = \frac{R_F}{R_1}(V_2 - V_1) \qquad \text{(Eq. 8-7)}$$

As before, resistor R_3 and R_4 are made equal for proper biasing; however, the additional restraint,

$$\frac{R_F}{R_1} = \frac{R_3}{2R_2} \qquad \text{(Eq. 8-8)}$$

must hold.

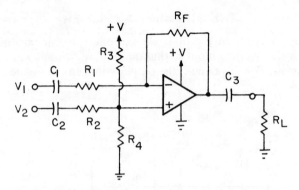

Fig. 8-5. Difference amplifier using a single supply voltage.

Example

Design an analog subtractor (i.e., $R_F/R_1 = 1$) having an input impedance of 10 kΩ, and a low-frequency response of 30 Hz for each input. The output drives a 1-kΩ load.

Since the gain is 1, resistors R_1, R_2, and R_F are all equal, and also equal to the input impedance which is 10 kΩ. From Equation 8-8, resistor R_3 (which also equals R_4) is computed:

$$\frac{R_F}{R_1} - \frac{R_3}{2R_2}$$

$$\frac{10 \text{ k}\Omega}{10 \text{ k}\Omega} = \frac{R_3}{(2)(10 \text{ k}\Omega)}$$

so that R_3 and R_4 are equal to 20 kΩ. The coupling capacitors, using Equations 8-3 and 8-4 are 0.5 μF, and 5 μF for C_1, C_2, and C_3, respectively. The completed design is shown in Fig. 8-6.

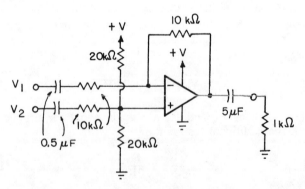

Fig. 8-6. An analog subtractor with an impedance of 10 kΩ for each input.

Before concluding this chapter, I would like to make one suggestion. If you ever consider the necessity of operating an ac op-amp circuit from a single supply voltage, you should strongly consider the circuits of the next chapter using *Norton op-amps*.

AN INTRODUCTION TO THE EXPERIMENTS

The experiments in this chapter are designed to demonstrate the operation of a few op-amp circuits using only a single supply voltage instead of the usual practice of using a dual or bipolar supply voltage. The experiments that you will perform can be summarized as follows:

Experiment No.	*Purpose*
1	Demonstrates the effect of varying the bias resistors on the quiescent dc output voltage.
2	Demonstrates the operation of an inverting amplifier, which is identical to the experiment in Chapter 2, except that we are using a single supply voltage.

EXPERIMENT NO. 1

Purpose

The purpose of this experiment is to demonstrate the effect of varying the bias resistors on the quiescent dc output voltage, using a type 741 op-amp powered by a single supply voltage.

Schematic Diagram of Circuit (Fig. 8-7)

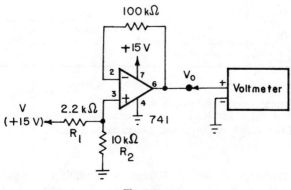

Fig. 8-7.

Design Basics

- Quiescent dc output voltage: $(V_o)_{dc} = \dfrac{R_2}{R_1 + R_2}(V)$

  Usually $(V_o)_{dc} = 0.5\ V$ $(R_1 = R_2)$

Step 1

Wire the circuit shown in the schematic diagram (Fig. 8-6). *Make sure that pin 4 of the 741 op-amp is connected to ground.*

Step 2

Apply power to the breadboard. First measure the supply voltage (V) with the voltmeter, recording your result below:

Supply voltage $(V) = $ _____ volts

Step 3

Next, measure the output voltage (V_o), while varying R_1 according the following table. Compare your measurements with the formula given in the *Design Basics* section. *Please, disconnect the power to the breadboard each time before changing R_1!*

R_1	V_o (measured)	V_o (calculated)
2.2 kΩ		
3.9 kΩ		
4.7 kΩ		
6.8 kΩ		
10 kΩ		
22 kΩ		
33 kΩ		
47 kΩ		

Step 4

At what value for R_1 is the quiescent dc output voltage equal to one-half the supply voltage you measured in Step 2?

The quiescent dc output voltage is one-half the supply voltage when R_1 equals R_2 (10 kΩ). In general, for proper operation from a single supply voltage, the quiescent dc output voltage is made equal to one-half the supply voltage.

EXPERIMENT NO. 2

Purpose

The purpose of this experiment is to demonstrate the operation of an inverting amplifier powered by a single supply voltage. This experiment is identical to the steps given in Experiment No. 3 of Chapter 2.

Schematic Diagram of Circuit (Fig. 8-8)

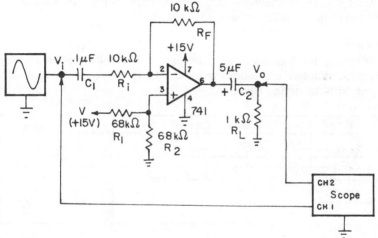

Fig. 8-8.

Design Basics

- Voltage gain: $\dfrac{V_o}{V_i} = -\dfrac{R_F}{R_i}$

- Quiescent dc output: $(V_0)_{dc} = \dfrac{R_2}{R_1 + R_2}(V)$

- Coupling capacitors: $C_1 = \dfrac{1}{2\pi f_c R_i}$

$$C_2 = \dfrac{1}{2\pi f_c R_L}$$

Step 1

Wire the circuit shown in the schematic diagram (Fig. 8-8).

Step 2

Set your oscilloscope for the following settings:

- Channels 1 & 2: 0.2 volt/division
- Time base: 1 msec/division
- DC coupling

Step 3

Apply power to the breadboard and adjust the generator output voltage at 0.6 volt peak-to-peak, and the frequency so that there are about 5 complete cycles for the 10 horizontal divisions (500 Hz). Is there any difference between the input and output signals?

The output signal is inverted as compared with the input signal, since we have an inverting amplifier. Otherwise the two amplitudes are the same since the voltage gain is 1.0.

Step 4

Keeping the input voltage constant at 0.6 volt, change resistor R_F and complete the following table. Do your experimental results agree with the design equation?

R_F	measured V_o (peak-to-peak)	voltage gain
27 kΩ		
39 kΩ		
47 kΩ		
82 kΩ		

CHAPTER 9

The Norton Op-Amp

INTRODUCTION

In this chapter, a different kind of op-amp will be presented. The Norton op-amp, or current-differencing amplifier, is designed to operate from a single ended supply. This amplifier makes use of a *current mirror* to achieve a noninverting input function. Application areas are similar to those discussed for the standard op-amp in the previous chapters.

OBJECTIVES

At the completion of this chapter, you will be able to do the following:
• Design and predict the performance of the following basic circuits using Norton op-amps:

an inverting amplifier
a noninverting amplifier
a summing amplifier
a difference amplifier
a voltage-controlled oscillator

OPERATION

The Norton op-amp, or *current-differencing amplifier,* is slightly different than the standard op-amp in its operation; even its symbol (Fig. 9-1) is different. The Norton op-amp operates on the *difference of the currents flowing into the inverting* $(-)$ *and noninverting* $(+)$

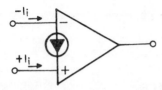

Fig. 9-1. Symbol for the Norton op-amp.

inputs. The best known Norton op-amp is the LM3900, made by National Semiconductor, whose pin configuration is shown in Fig. 9-2. You should notice that there are 4 independent amplifiers in this 14-pin package, which sells for about 50 cents in single quantities from several mail order houses. For about 13 cents an amplifier, you can hardly go wrong!

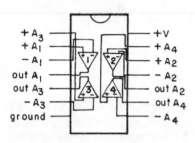

Fig. 9-2. LM3900 pin configuration.

Although it is of little importance here to discuss what makes this op-amp work, a very nice article covering Norton op-amps appeared in the June 1973 issue of *Popular Electronics.** In short, the input currents are differenced at the inverting input terminal, and this current difference then flows through an external feedback loop to produce the output voltage.

This type of op-amp can be used in most of the applications of the standard op-amp configurations that have been presented in this book, but using only a single supply voltage. Compared with the 741 op-amp, the major parameters are summarzed in Table 9-1.

Table 9-1. Major Parameters for 741 and 3900 Op-Amps

	741	3900
Supply voltage, V	±15	+4 to +36
Gain, open loop	200,000	2800
f_T, MHz	1.0	1.0
I_b, μA	80	30
Z_i, MΩ	2.0	1.0
Z_o, Ω	75	8 k
Slew rate, V/μsec	0.5	0.5

* Jung, W. G. "CDA—The New Current Differencing Amplifier." *Popular Electronics,* June, 1973, pp. 61-66.

BIASING

Since the Norton op-amp is to be powered from a single supply voltage, we must bias the device so that both positive and negative input signal excursions are amplified. This is similar to a *class-A* transistor or tube amplifier and the single supply biasing for standard op-amps (see Chapter 8). As shown in Fig. 9-3, the standard biasing technique is called *current-mirror biasing,* so that the *quiescent output voltage* (i.e., the dc output voltage with *no* input signal) *is nor-*

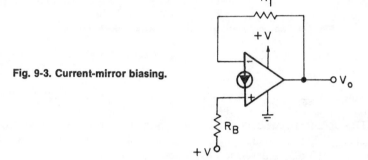

Fig. 9-3. Current-mirror biasing.

mally one-half the supply voltage. In practice the maximum output voltage that can be obtained from a LM3900 op-amp is approximately 0.7 volt less than the supply voltage, so that the quiescent voltage should be one-half this value. In terms of the resistors in the circuit of Fig. 9-3, the quiescent output voltage $(V_o)_{dc}$ is given by:

$$(V_o)_{dc} = \frac{R_1}{R_B} (V - 0.7) \qquad \text{(Eq. 9-1)}$$

In general, R_B is made twice the value of R_1. Unless stated otherwise, this method will be used for the majority of the Norton op-amp circuits.

THE NONINVERTING AMPLIFIER

The noninverting amplifier using the Norton op-amp is shown in Fig. 9-4, which is best suited as an *ac amplifier*. As with the standard op-amp, the input signal is applied to the noninverting input. Since almost all amplifier configurations using Norton op-amps are *ac amplifiers,* the input and output capacitors C_1 and C_2, respectively, are used to remove the dc component of the input and output signals.

Unlike the standard noninverting op-amp circuit, the voltage gain for the noninverting Norton op-amp circuit is given by:

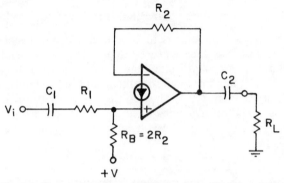

Fig. 9-4. The noninverting amplifier using a Norton op-amp.

$$\frac{V_o}{V_i} = \frac{R_2}{R_1} \qquad \text{(Eq. 9-2)}$$

In addition, the input impedance of this circuit is essentially equal to R_1.

The value for capacitor C_1 is determined by the low cutoff frequency (i.e., the frequency at which the voltage gain decreases by 0.707, or −3 dB as described in Chapter 8), so that:

$$(f_c)_{low} = \frac{1}{2\pi R_1 C_1} \qquad \text{(Eq. 9-3)}$$

Depending on the circuit's load resistance, C_2 is found in a similar manner,

$$f_c = \frac{1}{2\pi R_L C_2} \qquad \text{(Eq. 9-4)}$$

Example

Design a noninverting ac amplifier, using a Norton op-amp, with a low-frequency cutoff of approximately 30 Hz, a voltage gain of 10, and a 1-kΩ load.

Picking $R_1 = 100$ kΩ for example, then R_2 must equal 1 MΩ and R_B is $2R_2$, or 2 MΩ. Capacitor C_1 is found from Equation 9-3, so that:

$$C_1 = \frac{1}{2\pi f_c R_1}$$

$$= \frac{1}{(6.28)(30 \text{ Hz})(100 \text{ k}\Omega)}$$

$$\simeq 0.05 \ \mu F$$

Also, from Equation 9-4,

$$C_2 = \frac{1}{2\pi f_c R_L}$$

$$= \frac{1}{(6.28)(30 \text{ Hz})(1 \text{ k}\Omega)}$$

$$\simeq 5 \ \mu\text{F}$$

The final design is shown in Fig. 9-5.

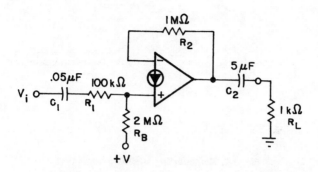

Fig. 9-5. A noninverting amplifier with a low-frequency cutoff of approximately 30 Hz and a voltage gain of 10.

THE INVERTING AMPLIFIER

As shown in Fig. 9-6, the Norton op-amp is wired as an inverting amplifier. The voltage gain is the same as for the noninverting Norton amplifier, except for the minus sign,

$$\frac{V_o}{V_i} = - \frac{R_2}{R_1} \qquad \text{(Eq. 9-5)}$$

The input impedance for this circuit is simply the value of R_1.

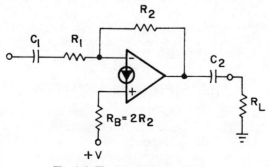

Fig. 9-6. The inverting amplifier.

SUMMING AMPLIFIERS

Depending on which input the input voltages are applied, we have either a noninverting summing amplifier (Fig. 9-7) or an inverting summing amplifier (Fig. 9-8). The output voltages for both circuits are the same, except for the polarity, which is:

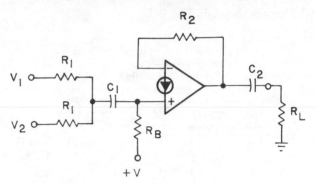

Fig. 9-7. Noninverting summing amplifier.

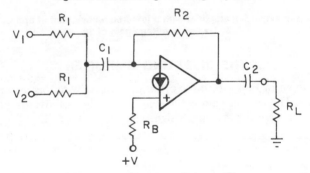

Fig. 9-8. Inverting summing amplifier.

$$V_o = \pm \frac{R_2}{R_1} (V_1 + V_2) \qquad \text{(Eq. 9-6)}$$

assuming that both input resistors are equal.

THE DIFFERENCE AMPLIFIER

As shown in Fig. 9-9, the Norton op-amp is used as a difference amplifier. The common-mode input impedance of each input is simply the value of the corresponding input resistor, while the differential impedance is the sum of the two. The output voltage, as a function of the two inputs, is:

$$V_o = \frac{R_3}{R_1} (V_2 - V_1) \qquad \text{(Eq. 9-7)}$$

Example

Using Fig. 9-9, design a difference amplifier with a voltage gain of 10, a low-frequency cutoff of approximately 30 Hz, an input impedance of 10 kΩ for both inputs, and a 1-kΩ load.

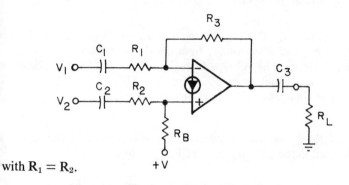

with $R_1 = R_2$.

Fig. 9-9. The difference amplifier.

Since R_1 and R_2 must be 100 kΩ each (the input impedance for each input, right?), then R_3 is 1 MΩ. In addition, the bias setting resistor R_B is $2R_3$, or 2 MΩ.

Capacitors C_1, C_2, and C_3 are found from Equations 9-3 and 9-4. As demonstrated in the previous example, $C_1 = C_2 = 0.05 \ \mu F$, and $C_3 = 5 \ \mu F$. The completed design is shown in Fig. 9-10.

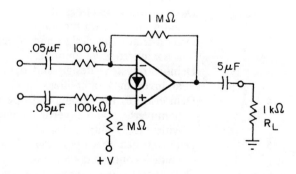

Fig. 9-10. Difference amplifier with a voltage gain of 10 and an input impedance of 100 kΩ.

OTHER APPLICATIONS

As has been shown in this book, there are many other functions possible using standard op-amps, and the Norton op-amp is no exception. These include inverting and noninverting comparators, window comparators, function generators, logic gates, active filters, regulators, and phase-locked loops. The following references describe some of the many applications in addition to the basic configurations that were already presented in this chapter.

1. Jung, W. G. "CDA—*The New Current Differencing Amplifier.*" *Popular Electronics,* June, 1973, pp. 61-66.
2. Jung, W. G. *The IC Op-Amp Cookbook.* Howard W. Sams & Co., Inc., Indianapolis, 1974, pp. 474-514.
3. *Linear Data Book,* National Semiconductor.

Several of these applications which are not described in this chapter will be found in the *Experiments* section to follow.

AN INTRODUCTION TO THE EXPERIMENTS

The following experiments in this chapter are designed to demonstrate the design and operation of several circuits using the LM3900 Norton op-amp with a single power supply.

The experiments that you will perform can be summarized as follows:

Experiment No.	Purpose
1	Demonstrates the effect of varying the bias setting resistor on the output voltage.
2	Demonstrates the design and operation of a noninverting ac amplifier.
3	Demonstrates the design and operation of an inverting ac amplifier.
4	Demonstrates the design and operation of noninverting and inverting comparators with a LED output indicator.
5	Demonstrates the operation of a simple voltage controlled oscillator.
6	Demonstrates the operation of several digital logic gates with a LED output indicator.

EXPERIMENT NO. 1

Purpose

The purpose of this experiment is to demonstrate the effect of varying the bias setting resistor upon the quiescent dc output voltage, using a LM3900 Norton op-amp.

Pin Configuration of LM3900 Norton Op-Amp (Fig. 9-11)

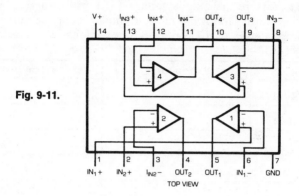

Fig. 9-11.

Schematic Diagram of Circuit (Fig. 9-12)

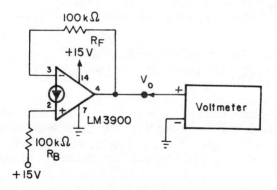

Fig. 9-12.

Design Basics

- Quiescent dc output voltage. $V_o = \dfrac{R_F}{R_B} \, (V)$

    Usually, $R_B = 2R_F$

Step 1

Wire the circuit shown in the schematic diagram (Fig. 9-12), and then apply power to the breadboard. First, measure the power supply voltage (V) with the voltmeter, recording your result below:

$$V_{(power\ supply)} = \underline{\hspace{2cm}} volts$$

Step 2

Next measure the output voltage (V_o) while varying resistor R_B according to the following table. Compare your measurements with the formula given in the *Design Basics* section.

R_B	V_o (measured)	V_o (calculated)
100 kΩ		
150 kΩ		
200 kΩ		
270 kΩ		
330 kΩ		
390 kΩ		
470 kΩ		

Step 3

At what value for the resistor R_B is the quiescent output voltage about one-half the supply voltage you determined in Step 1?

The quiescent output voltage is one-half the supply voltage when R_B is about 200 kΩ, or twice the value of the feedback resistor R_F. When R_B was equal to 100 kΩ, did you notice that the measured output voltage was about 0.8 volt less than the supply voltage?

The quiescent output voltage with R_B equal to R_F should be approximately 0.7 to 0.8 volt less than the supply voltage that you measured in Step 1, and is the maximum output voltage that can be obtained from this op-amp. This maximum output voltage is often referred to as the *saturated output voltage*, V_{SAT}. In reality, it is this saturated output voltage that should be used in place of the supply voltage in the design equation.

When I performed this experiment, my measurements were as follows:

$$V_{supply} = \underline{+14.38} \text{ volts}$$
$$V_{SAT} = \underline{+13.59} \text{ volts} \quad (R_B = 100 \text{ k}\Omega)$$
$$V_o = \underline{+7.28} \text{ volts} \quad (R_B = 200 \text{ k}\Omega)$$

However, when I tried this experiment with 17 other type LM3900 op-amps, I noticed that the quiescent output voltage varied from 7.22 to 7.29 volts ($\overline{x} = 7.40$, $\pm SD = 0.16$) with R_B equal to 200 kΩ. The only explanation that can be offered for this behavior is that there are slight differences between op-amps of the same type. Although, in general, R_B should be twice that of R_F, resistor R_B may have to be experimentally determined to give the best results.

EXPERIMENT NO. 2

Purpose

The purpose of this experiment is to demonstrate the operation of a noninverting ac amplifier, using a type LM3900 op-amp with a single supply voltage.

Schematic Diagram of Circuit (Fig. 9-13)

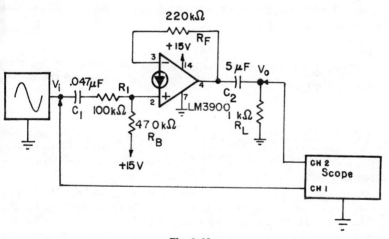

Fig. 9-13.

Design Basics

- Voltage gain: $\dfrac{V_o}{V_i} = \dfrac{R_F}{R_1}$

- Quiescent dc output: $(V_o)_{dc} = \dfrac{R_F}{R_B} \, (V)$

- Coupling capacitors: $C_1 = \dfrac{1}{2\pi f_c R_1}$

$$C_2 = \dfrac{1}{2\pi f_c R_L}$$

Step 1

Wire the circuit shown in the schematic diagram (Fig. 9-13).

Step 2

Set your oscilloscope for the following settings:
- Channels 1 & 2: 0.2 volt/division
- Time base: 1 msec/division
- DC coupling

Step 3

Apply power to the breadboard and adjust the generator's output voltage at 0.2 volt peak-to-peak and the frequency at 400 Hz (4 complete cycles). What is the difference between the input and output signals?

The only difference between these two signals is that the output signal is larger than the input, since both signals are in phase as this is a noninverting amplifier.

Step 4

Measure the peak-to-peak output voltage and determine the voltage gain. How does it compare with the equation given in the *Design Basics* section?

You should have measured an output voltage of approximately 0.44 volt peak-to-peak, giving a voltage gain of 2.20.

Step 5

Verify the design equation for the circuit's voltage gain by changing the value of resistor R_F. Don't forget to also change resistor R_B so that it is approximately twice that of R_F!

EXPERIMENT NO. 3

Purpose

The purpose of this experiment is to demonstrate the design of an inverting ac amplifier, using a type LM3900 Norton op-amp. In ad dition, the voltage gain for this circuit is controlled by using a type 4016 CMOS analog switch.

Pin Configuration of Integrated Circuit Chips (Fig. 9-14)

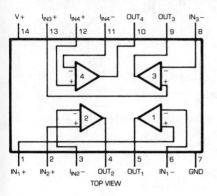

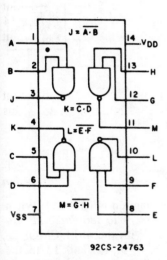

CD4011A

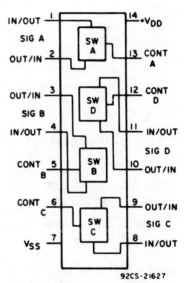

Functional Diagram

Fig. 9-14.

Schematic Diagram of Circuit (Fig. 9-15)

Design Basics

- Voltage gain: $\dfrac{V_o}{V_i} = -\dfrac{R_F}{R_i}$

- Quiescent dc output: $(V_o)_{dc} = \dfrac{R_F}{R_B} \ (V)$

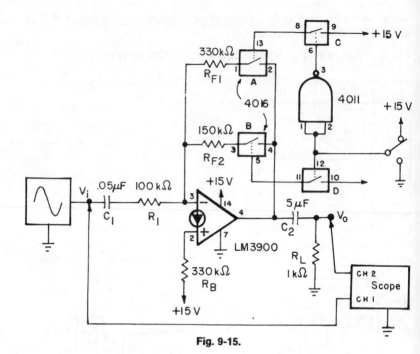

Fig. 9-15.

- Coupling capacitors: $C_1 = \dfrac{1}{2\pi f_c R_1}$

$$C_2 = \dfrac{1}{2\pi f_c R_L}$$

Step 1

Wire the circuit shown in the schematic diagram (Fig. 9-15). When using CMOS integrated circuits, *it is necessary to terminate all unused inputs either at logic 0 or at logic 1*. With the 4011 CMOS 2-input quad NAND gate, connect pins 5, 6, 8, 9, 12, and 13 to the +15-volt supply. Also, don't forget the 4011 power supply connections: pin 14 (+15 volts) and pin 7 (ground), as these power connections are usually omitted from schematic diagrams.

For the 4016 CMOS analog switch, connect pin 14 to +15 volts, and pin 7 to ground.

Step 2

Set your oscilloscope for the following settings:
- Channels 1 & 2: 0.2 volt/division
- Time base: 1 msec/division
- DC coupling

Step 3

Switch the spdt switch to +15 volts and then apply power to the breadboard. Adjust the peak-to-peak input voltage at 0.4 volt (2 divisions) and the frequency at 500 Hz (5 complete cycles).

Step 4

Measure the peak-to-peak output voltage, and then determine the voltage gain for this configuration. From your result, which feedback resistor is the one that is "switched" into the circuit?

You should have measured a peak-to-peak output voltage of approximately 0.6 volt, giving a voltage gain of 1.50. Consequently, from the design equation, you should have concluded that the feedback resistor is the 150-kΩ resistor. In addition, the output is *inverted* with respect to the input, since this circuit is an *inverting amplifier*.

Step 5

Now switch the spdt switch to ground. What now is the voltage gain?

The output voltage should be approximately 1.32 volts, giving a voltage gain of 3.30. The 330-kΩ resistor is now switched into place as the feedback resistor.

This circuit is an interesting one, in that we are now able to alter the voltage gain of this circuit by using analog switches. The switches A and B permit one to change the value of the feedback resistor from 150 kΩ to 330 kΩ. Switches A and B essentially function as a pair of spst switches, with only one closed at any given time. Switches C and D with the 4011 NAND gate is a configuration that is equivalent to an spdt switch. When the *manual* spdt switch is connected to +15 volts, switch D is *closed,* and switch C is *open.* When the manual spdt switch is grounded, switch D is now open and C is closed. Instead of using a manual switch, a suitable digital circuit, such as a decoder or a comparator (see Experiment No. 4) can be used to control the op-amp's gain at a specific time.

EXPERIMENT NO. 4

Purpose

The purpose of this experiment is to demonstrate the operation of noninverting and inverting comparators, using a type LM3900 Norton op-amp with a LED monitor.

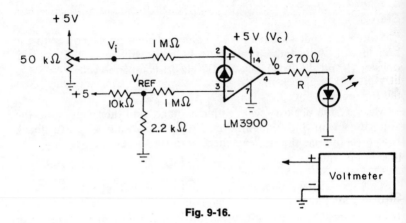

Fig. 9-16.

Schematic Diagram of Circuit (Fig. 9-16)

Design Basics

- When: $V_i < V_{REF}$, $V_o \simeq 0$ (LED unlit)

   $V_i > V_{REF}$, $V_o \simeq 4.2V$ (LED lit)

   using a +5-volt supply.

Step 1

If you are using a +5-volt supply, wire the circuit as shown in the schematic diagram (Fig. 9-16). If a +15-volt supply is used, use a 1.5-kΩ resistor for R. For other supply voltages between 4 and 36 volts, calculate R from the formula:

$$R = \frac{V_c - 2.6}{10 \text{ mA}}$$

Step 2

Apply power to the breadboard, and with a voltmeter (preferably a digital voltmeter, or dvm), accurately measure V_{REF}, recording your result below:

$$V_{REF} = \text{_____} \text{ volts}$$

Using a +5-volt supply, I measured +0.95 volt; for a +15-volt supply, I measured +2.78 volts.

Step 3

Now connect the voltmeter to measure the voltage V_i. Vary the 50-kΩ potentiometer so that V_i is *less than* the value of V_{REF} that you measured in Step 2. Is the LED monitor lit or unlit?

The LED monitor should be *unlit,* since the input voltage is less than the reference voltage.

Step 4

Now slowly increase the input voltage until the LED monitor is lit at its full brightness. Is the input voltage less than or greater than the reference voltage that you measured in Step 2?

The input voltage should be *greater than* the reference voltage. Is this an inverting or a noninverting comparator?

This is a *noninverting* comparator.

Step 5

Disconnect the power from the breadboard and then reverse the op-amp's + and − input connections, forming an *inverting* comparator.

Step 6

Apply power to the breadboard and vary V_i so that it is *less than* V_{REF} (Step 2). Is the LED monitor lit or unlit?

The LED monitor should now be *lit,* since we now have an inverting comparator.

Step 7

Now slowly increase the input voltage until the LED monitor is unlit. Is the input voltage less than or greater than the reference voltage you measured in Step 2?

The input voltage should be greater than the reference voltage.

Step 8

Repeat this experiment for a different value of V_{REF} by changing the value of the 2.2-kΩ resistor to 10 kΩ, for example.

EXPERIMENT NO. 5

Purpose

The purpose of this experiment is to demonstrate the operation of a voltage-controlled oscillator, made from a LM3900 Norton op-amp.

Schematic Diagram of Circuit (Fig. 9-17)

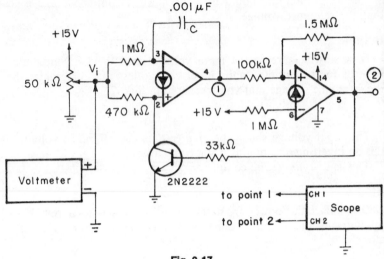

Fig. 9-17.

Step 1

Set your oscilloscope for the following settings:
- Channel 1: 0.5 volt/division
- Channel 2: 5 volts/division
- Time base: 1 msec/division
- DC coupling

In addition, a voltmeter, preferably a digital type, capable of measuring 0–10 volts will be required.

Step 2

Wire the circuit shown in the schematic diagram (Fig. 9-17). Apply power to the breadboard and adjust the 50-kΩ potentiometer so that the input voltage V_i is approximately 1.50 volts. What do you see on the oscilloscope's display?

On Channel 1 you should see a triangle-shaped waveform, and on Channel 2 you should see a square wave.

Step 3

Vary the potentiometer to increase the input voltage. Does the frequency of these two waveforms increase or decrease?

With increasing input voltage, the frequency will *increase*.

Step 4

Now adjust the potentiometer so that 1 complete cycle occupies the 10 horizontal divisions of the oscilloscope display (i.e., 100 Hz). Vary the input voltage so that there will be an integer number of complete cycles on the display in order to complete the following table. In addition, graph your results on the graph (Fig. 9-18) provided for this purpose.

Number of cycles/ 10 divisions	Frequency Hz	V_i Volts
1	100	
2	200	
3	300	
4	400	
5	500	
6	600	
7	700	
8	800	
9	900	
10	1,000	

You should find that the frequency of the voltage controlled oscillator increases *linearly* with increasing input voltage. When I performed this experiment, I found that the output frequency increased 56 Hz for every 0.1 volt increase.

EXPERIMENT NO. 6

Purpose

The purpose of this experiment is to demonstrate that the Norton op-amp can be connected to function as the following logic elements: 2-input AND, NAND, OR, and NOR gates, and an inverter. You will experimentally determine the truth table for these logic functions by observing the status of the LED monitor.

Schematic Diagram of Circuit (Fig. 9-19)

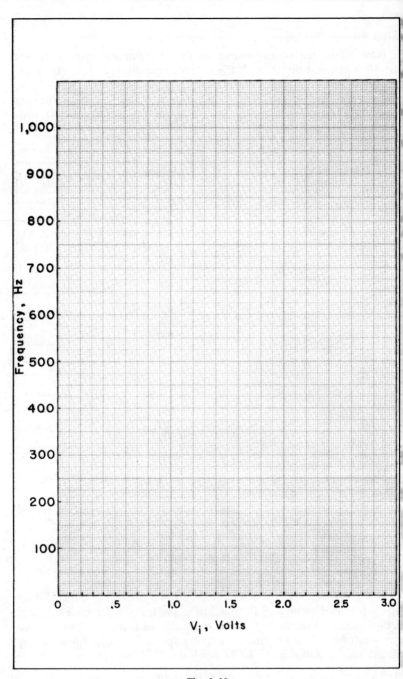

Fig. 9-18.

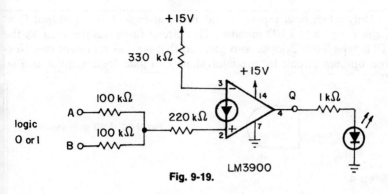

Fig. 9-19.

Step 1

Wire the circuit shown in the schematic diagram (Fig. 9-19). Initially, *ground both inputs* (A and B), which corresponds to a *logic 0*. A *logic 1* input corresponds to a connection to the +15-volt supply. The above circuit is a *2-input* AND *gate.*

Step 2

Apply power to the breadboard. The LED monitor should be *un-lit,* indicating that the output (Q) from this 2-input AND gate is at logic 0 (a logic 1 implies a lit LED), when both inputs are at logic 0.

Step 3

By varying the input logic settings, complete the following *truth table* for this 2-input AND gate.

Inputs		Output
A	**B**	**Q**
0	0	0
1	0	
0	1	
1	1	

If you have done everything correctly, your results should be identical to the 2-input AND gate truth table shown below:

Inputs		Output
A	**B**	**Q**
0	0	0
1	0	0
0	1	0
1	1	1

Only when *both* inputs A and B are at *logic 1* is the output Q at *logic 1*, i.e., a lit LED monitor. This circuit functions the same as the TTL type 7408 2-input AND gate, so that we can represent this Norton op-amp circuit by the equivalent AND gate logic symbol in Fig. 9-20.

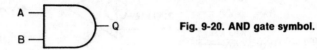

Fig. 9-20. AND gate symbol.

Step 4

Disconnect the power from the breadboard and *reverse the connections to the op-amp inverting (−) and noninverting (+) inputs* (pins 3 and 2, respectively). This connection now results in a 2-*input NAND gate,* which is the opposite of a 2-input AND gate. In addition, set both inputs A and B at logic 0.

Step 5

Apply power to the breadboard. The LED monitor should be lit, indicating that the output from this 2-input NAND gate is at logic 1 when both inputs are at logic 0.

Step 6

By varying the input logic settings, complete the following truth table for this 2-input NAND gate:

Inputs		Output
A	B	Q
0	0	1
1	0	
0	1	
1	1	

Your results should be the same as the following truth table:

Inputs		Output
A	B	Q
0	0	1
1	0	1
0	1	1
1	1	0

Only when both inputs are at *logic 1* is the output at *logic 0*. Compare this truth table with the one in Step 3. The outputs are opposite for the same logic inputs. This circuit functions the same as the TTL type 7400 2-input NAND gate, so that we can represent this Norton op-amp circuit by the equivalent NAND gate logic symbol in Fig. 9-21.

Fig. 9-21. NAND gate symbol.

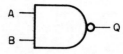

Step 7

Disconnect the power from the breadboard. Now wire the circuit shown in the schematic diagram in Fig. 9-22, which corresponds to a *2-input* OR *gate:*

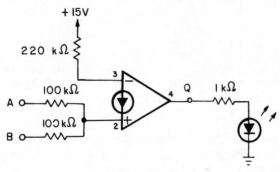

Fig. 9-22.

Step 8

Initially, set both inputs at logic 0. Apply power to the breadboard. The LED monitor should be unlit, or logic 0.

Step 9

By varying the input logic setting, complete the following truth table for this 2-input OR gate:

Inputs		Output
A	B	Q
0	0	0
1	0	
0	1	
1	1	

197

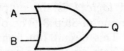

Fig. 9-23. OR gate symbol.

Your results should be the same as the following truth table:

Inputs		Output
A	B	Q
0	0	0
1	0	1
0	1	1
1	1	1

Only when both inputs are at *logic 0* is the output at *logic 0*. This circuit functions the same as the TTL type 7432 2-input OR gate, so that we can represent this Norton op-amp circuit by the equivalent OR gate logic symbol in Fig. 9-23.

Step 10

Again disconnect the power from the breadboard and *reverse the op-amp input connections,* as in Step 4. The new connection results in a *2-input* NOR *gate,* which is the opposite of a 2-input OR gate. In addition, set both inputs at logic 0.

Step 11

Apply power to the breadboard. The LED monitor should be lit. By varying the input logic settings, complete the following truth table for this 2-input NOR gate:

Inputs		Output
A	B	Q
0	0	1
1	0	
0	1	
1	1	

Your results should be the same as the following truth table:

Inputs		Output
A	B	Q
0	0	1
1	0	0
0	1	0
1	1	0

Only when both inputs are at *logic 0* is the output a *logic 1*. Compare this table with the one in Step 9. The outputs are opposite for the same logic input. This circuit functions the same as the TTL type 7402 2-input NOR gate, so that we can represent this Norton op-amp circuit by the equivalent NOR gate logic symbol in Fig. 9-24.

Fig. 9-24. NOR gate logic symbol.

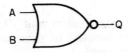

Step 12

Disconnect the power from the breadboard. Now wire the circuit shown in the schematic diagram in Fig. 9-25, which is an *inverter*.

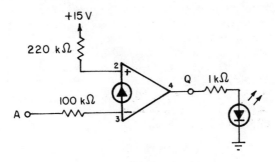

Fig. 9-25. Inverter circuit.

Step 13

Initially set the single input A at logic 0. Apply power to the breadboard. The LED monitor should be lit.

Step 14

Now set the input A at logic 1. Is the LED monitor lit or unlit?

The LED monitor should be *unlit,* since the output of the inverter is the *opposite* of its input, as shown by the following truth table:

Input	Output
A	Q
0	1
1	0

This circuit functions the same as the TTL type 7404 inverter, so that we can represent this Norton op-amp circuit by the equivalent inverter logic symbol in Fig. 9-26.

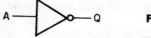

Fig. 9-26. Inverter logic symbol.

If you have worked with TTL integrated circuits before, you may have been tempted to determine what happens when any input is left unconnected.

An unconnected input to a TTL integrated circuit gate is generally logic 1. However, with the Norton op-amp logic circuits, the inputs must be either at logic 0 or logic 1; *they should not be left unconnected.*

CHAPTER 10

The Instrumentation Amplifier

INTRODUCTION

In this final chapter, a different type of integrated circuit amplifier will be described. The instrumentation amplifier, also called the IA, is a closed-loop, differential-input device, and differs fundamentally from the standard op-amp.

OBJECTIVES

At the completion of this chapter, you will be able to do the following:
- Explain the difference in performance between IAs and standard op-amps.
- Test the characteristics of the Burr-Brown 3660J instrumentation amplifier.

THE INSTRUMENTATION AMPLIFIER

The instrumentation amplifier, or IA, is a "committed-gain" amplifier, with internal high-precision feedback network. From Chapter 1, the standard op-amp is an *open-loop uncommitted device* whose closed-loop performance depends on the external networks used.*

---

* Two well written references on instrumentation amplifiers are:
1. Miller, B. "For Tough Measurements, Try IA's." *Electronic Design,* August 2, 1974, pp. 84-87.
2. Goodenough, F. "Op Amp or Instrumentation Amp: When Do You Use Which and Why?" *EDN,* May 20, 1977, pp. 109-113.

As pointed out in Chapter 2, parameters such as input and output impedance, and output offset voltage are dependent upon the external closed-loop connections. In addition, the standard op-amp circuit usually involves some design tradeoffs when it is necessary to amplify only *low-level* signals, usually a few microvolts to a few millivolts, in the presence of high common-mode voltages (noise).

The IA is used almost always as a difference amplifier, so that it amplifies the difference between two ground-referenced input signals. One outstanding feature of the IA is its high *common-mode rejection* (CMR), or its ability to reject ground-referenced in-phase signals common to both inputs, such as 60-Hz hum, even though these two input signal sources may be unbalanced by as much as 1 kΩ.

As shown in Fig. 10-1, a standard op-amp is used as a difference amplifier.

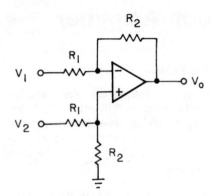

Fig. 10-1. Op-amp difference amplifier.

Ideally, for low-level signals, we would like to have high gain and high input impedance. From Chapter 2, the output voltage for this circuit is ideally,

$$V_o = \frac{R_2}{R_1} (V_2 - V_1) \qquad \text{(Eq. 10-1)}$$

In a practical situation, we would naturally expect that there will also be a voltage signal (V_{CM}), such as 60-Hz noise, that is simultaneously present at both inputs. Consequently, this *common-mode* voltage adds an "error" term to Equation 10-1, so that:

$$V_o = \frac{R_2}{R_1} (V_2 - V_1) + \frac{R_2}{R_1} \left(\frac{V_{CM}}{CMRR} \right) \qquad \text{(Eq. 10-2)}$$

where CMRR is the IA's common-mode rejection ratio, expressed in dB (see Chapter 1). For Equation 10-2 to approach Equation 10-1, the CMRR must be large enough to make this error term negligible.

In order to simultaneously have high gain and high input impedance, we have conflicting design constraints. For this circuit to have high gain, the input resistor R_1 must be small. On the other hand, R_1 must be large for high input impedance. The beauty of the IA is that it eliminates most of the above problems of using standard op-amps, particularly when used as a difference amplifier. The IA responds only to the difference between the two input signals, and it also has an extremely high input impedance from each input to ground. IAs use a single resistor that sets the gain *which does not affect the input impedance, but only the CMRR.* We now can have our cake and eat it too!

So that we can take a closer look at IAs, we will consider the type 3660J IA, manufactured by Burr-Brown, as a representative example. The 3660J is a 10-pin integrated circuit in a T0-100 style can, as shown in Fig. 10-2. It *is not cheap!* This model presently sells for about \$11 in single quantities. If nothing else, this price should make you respect the fact that this is a "precision" device, compared with its distant cousin, the 741 op-amp, selling for about 3/\$1.

Using the 3660J IA, the output voltage is given by,

$$V_o = [V_2 - V_1]\frac{100\text{ k}\Omega}{R_G}$$

for the circuit of Fig. 10-3. Resistor R_G is the external gain-setting resistor. Like the standard op-amp, IAs operate from a bipolar, or dual supply, whose connections are usually omitted from schematic diagrams.

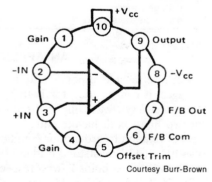

Fig. 10-2. The Burr-Brown 3660J instrumentation amplifier.

Courtesy Burr-Brown

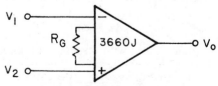

Fig. 10-3. IA difference amplifier.

As summarized from its data sheet, the specifications for the 3660J IA are listed below in Table 10-1.

Table 10-1. 3660J IA Specifications

Voltage gain (G)	
range	1–1000 (0 to 60 dB)
Output	
rated output voltage	$\pm$10 V
rated output current	$\pm$10 mA
dc output impedance (@ G = 100)	0.15 Ω
Input	
differential impedance	2×10^6 Ω/G
common-mode impedance	2×10^6 Ω/G
input voltage range	$\pm$10 V
CMRR, dc to 60 Hz @ G = 10,	
1 kΩ unbalance	76 dB
@ G = 1000, 1 kΩ unbalance	96 dB (minimum)
Offsets	
offset voltage (referred to input)	$\pm \left(6 + \dfrac{600}{G} \right)$ mV
offset voltage (referred to output)	$\pm \left(6 + \dfrac{600}{G} \right)$ G mV
bias current (each input)	200 nA
Dynamic response (@ G = 100)	
for $\pm$1% minimum flatness	10 kHz (typical)
slew rate	1.8 V/μsec
Power supply	
rated voltage	$\pm$15 V
voltage range	$\pm$7 to 20 V
max quiescent current	$\pm$6 mA

AN APPLICATION

Because the IA responds only to the difference between two input voltages, it can be used for both *balanced* and *unbalanced* systems. A *balanced system* is one that the output of a signal source appears on two lines, both having equal source impedances and output voltages in relation to ground, or common-mode level, such as the output of a balanced Wheatstone bridge (Fig. 10-4).

The voltage at point 1, with respect to ground, is:

$$V_1 = \left(\frac{R_A}{R_A + R_B} \right) V \qquad \text{(Eq. 10-4)}$$

and the voltage at point 2, also with respect to ground, is:

$$V_2 = \left(\frac{R_X}{R_C + R_X} \right) V \qquad \text{(Eq. 10-5)}$$

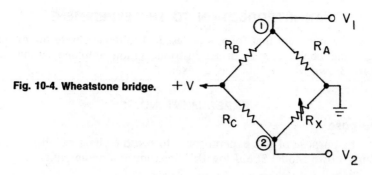

Fig. 10-4. Wheatstone bridge.

When this bridge is "balanced," (the bridge output voltage $V_1 - V_2$ is zero), Equations 10-4 and 10-5 are equal to each other, so that:

$$\frac{R_A}{R_A + R_B} = \frac{R_X}{R_C + R_X}$$

$$R_A(R_C + R_X) = R_X(R_A + R_B)$$

or,

$$R_X = \frac{R_A + R_C}{R_B} \qquad \text{(Eq. 10-6)}$$

The Wheatstone bridge can then be used either as a balanced bridge, where the unknown resistance R_X is measured by adjusting the value of the other bridge resistors for zero output, or as an unbalanced bridge where the unknown resistance is measured by measuring the output voltage of the bridge. *The balanced bridge is only used in static measurement situations. However the unbalanced bridge is used in both static and dynamic measurement situations.* Most transducer (e.g., a strain gauge) applications use the Wheatstone bridge in its unbalanced condition, which is ideal for the IA (Fig. 10-5), since it is specifically designed to reject the common-mode voltage present at terminals 1 and 2 of the bridge.

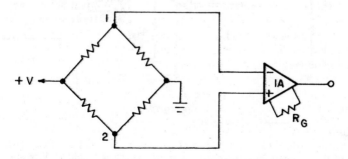

Fig. 10-5. Wheatstone bridge used with instrumentation amplifier.

AN INTRODUCTION TO THE EXPERIMENT

The experiment that follows is designed to demonstrate the operation and characteristics of an instrumentation amplifier, using the Burr-Brown Model 3660J as an example.

EXPERIMENT NO. 1

Purpose

The purpose of this experiment is to test the characteristics of the Burr-Brown Model 3660J low-drift instrumentation amplifier.*

Pin Configuration of the 3660J Instrumentation Amplifier (Fig. 10-6)

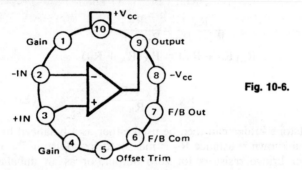

Fig. 10-6.

Schematic Diagram of Circuit (Fig. 10-7)

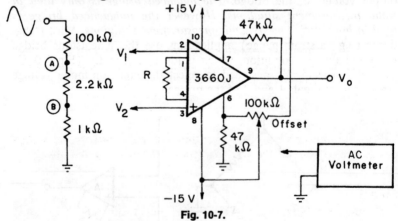

Fig. 10-7.

---

* This experiment is attributed to Mr. David Larsen and Dr. Peter Rony, as part of their course, *Electronics for Scientists,* which is taught at Virginia Polytechnic Institute and State University, Blacksburg, Virginia.

Design Basics

- Output voltage: $V_o = (V_2 - V_1) \dfrac{100 \text{ k}\Omega}{R}$

Step 1

The first point that should be emphasized is that the Model 3660J is a relatively expensive linear integrated circuit. If you purchase a single device, you will have to pay approximately $11.00. Therefore, I urge you to be very careful when you use it.

Step 2

Study the pin configuration of the device. Note that it is housed in a 10-lead, TO-100 style metal can. As an aid for the proper insertion of the integrated circuit into the breadboard, you can use the AB-10 Adapter Board, sold by E&L Instruments, Inc., shown in Fig. 10-8.

Fig. 10-8. E & L Instruments AB-10 Adapter Board.

Courtesy E & L Instruments, Inc.

This device allows direct insertion of the 3660J IA into the breadboard sockets without having to bend the leads into the 0.1″ dual-inline (DIP) spacing. If you do not use this aid, be very careful when bending the leads and inserting the IA into the breadboard!

Step 3

With the power supply off, wire the +15 volts and the −15 volts power connections to the IA. Connect −15 volts to pin 8 and +15 volts to pin 10.

Step 4

Connect a 1-kΩ resistor (R) between pins 1 and 4.

Step 5

Conect a 47-kΩ resistor between pins 7 and 9, and another 47-kΩ resistor between pin 6 and ground. These two resistors do not affect the grain of the device; they are feedback resistors that improve the performance characteristics of the Model 3660J device.

Step 6

Next, ground both inputs of the IA (pins 2 and 3). Now apply power to the breadboard and measure the output voltage at pin 9 with a voltmeter, preferably a digital type.

When this experiment was performed, I measured an offset voltage of -0.458 volts. This will depend on the setting of the 10-kΩ "offset" potentiometer. In most cases, you will probably have a dc offset voltage. Vary the "offset" potentiometer until the output voltage is zero, to within 1 or 2 mV.

Step 7

Using the resistance voltage divider, shown at the far left of the schematic diagram (Fig. 10-7), connect a sine-wave voltage source, such as a 6.3-volt filament transformer, which has an input frequency of 60 Hz. In my experiment, I used a sine-wave generator whose output voltage (V_1) was measured to be 0.903 volts rms.

Connect the noninverting input V_2 (pin 3) to point A on the resistance divider, and the inverting input V_1 (pin 2) to point B. Now, with an ac voltmeter, measure the voltage at points A and B, recording your values below:

$$V_A(V_2) = \underline{\hspace{1cm}} \text{ mV rms}$$

$$V_B(V_1) = \underline{\hspace{1cm}} \text{ mV rms}$$

There exist two different ways in which you can connect the differential inputs V_1 and V_2 to the voltage divider circuit:

Case 1: V_2 is greater than V_1
Case 2: V_1 is greater than V_2

The measurements that you have just made in this step correspond to Case 1.

Step 8

Now measure the output voltage, recording your result below:

$$V_o = \underline{\hspace{1cm}} \text{ mV rms}$$

Determine the *differential gain, A_d,* by dividing the output voltage by the differential input voltage $V_2 - V_1$, recording your result below:

$$A_d = \underline{\hspace{1cm}}$$

From the formula given in the *Design Basics* section, when a 1 kΩ resistor is used for the gain-setting resistor R, the differential gain is 100 kΩ/1 kΩ, or 100. How close were you?

Step 9

Now reverse the inputs, connecting V_1 to point A, and V_2 to point B (Case 2), and repeat Steps 7 and 8. You should find that the output voltage is negative, since V_1 is greater than V_2. Otherwise, the differential gain should approximately be the same. In my case I measured a differential gain of 93.3.

Step 10

Now connect both inputs to point V_i (the sine-wave input voltage). Measure this voltage, and record your result below, which is the *common-mode input voltage*, $V_i(cm)$:

$$V_i(cm) = \text{_____} \text{ volts rms}$$

Step 11

Measure the output voltage, which is now called the common-mode output voltage, $V_o(cm)$, and record your result below:

$$V_o(cm) = \text{_____} \text{ mV rms}$$

Step 12

The ratio of the common-mode output voltage to the common-mode input voltage is the common-mode gain, A_{CM}. From the values of Steps 10 and 11, determine the common-mode gain, recording your result below:

$$A_{CM} = \text{_____}$$

In my experiment, when $V_i(cm)$ was 6.24 volts rms, the corresponding output voltage was 7.4 mV rms, giving a common-mode gain of 0.00119.

Step 13

Now determine the common-mode rejection ratio in dB, according to the formula,

$$CMRR = 20 \log_{10} \frac{A_d}{A_{CM}}$$

using the differential gain, as determined in Step 8.

$$CMRR = \text{_____} \text{ dB}$$

For the 3660J IA the CMRR (typical) for a differential gain of 1000 is 86 dB. For my experiment, I determined a CMRR of 97.1 dB.

Step 14

Momentarily disconnect the power from the breadboard and replace the gain setting resistor R by a 100 Ω resistor. The theoretical voltage gain is now 1000. Repeat Steps 7 through 13 based on a differential gain of 1000. Do your results reflect this value? Is the CMRR for a differential gain of 1000 different than for a differential gain of 100?

When I performed this experiment, I determined a differential gain of 972. For this gain, I determined a CMRR of 104.2 dB, which was greater than the value determined for a differential gain of 100. According to the manufacturer's data sheet, the typical value for the CMRR at a differential gain of 1000 is 96 dB. For the 3660J IA, the CMRR increases as the differential gain is increased. My results are consistent with the data sheet, as I hope that yours were also.

References

BOOKS

1. Berlin, H. M. *The Design of Active Filters, With Experiments.* Indianapolis: Howard W. Sams & Co., Inc., 1978.

2. Hoenig, S., and L. Payne. *How to Build and Use Electronic Devices Without Frustration, Panic, Mountains of Money, or an Engineering Degree.* Boston: Little, Brown and Co., 1973.

3. Jung, W. G. *IC Op-Amp Cookbook.* Indianapolis: Howard W. Sams & Co., Inc., 1974.

4. Lenk, J. D. *Manual for Operational Amplifier Users.* Reston: Reston Publishing Co., 1976.

5. Melen, R., and H. Garland. *Understanding IC Operational Amplifiers.* Indianapolis: Howard W. Sams & Co., Inc., 1971.

6. Prensky, S. D. *Manual of Linear Integrated Circuits.* Reston: Reston Publishing Co., 1974.

7. Tobey, G. E., Graeme, J. G., and L. P. Huelsman. *Operational Amplifiers— Design and Applications.* New York: McGraw-Hill, 1971.

SHORT ARTICLES

1. Berlin, H. M. "Design Your Own Active Audio Filters." *QST,* June, 1977, pp. 32-34.

2. Goodenough, F. "Op Amp or Instrumentation Amp: When Do You Use Which and Why?" *EDN,* May 20, 1977, pp. 109-113.

3. Hart, B. "How Those Triangle Things Work." *73,* June, 1976. pp. 60-65.

4. Jung, W. G. CDA—"The New Current Differencing Amplifier." *Popular Electronics,* June, 1973, pp. 61-66.

5. Miller, B. "For Tough Measurements, Try IA's." *Electronic Design,* August 2, 1974, pp. 84-87.

6. Sheingold, D., and F. Pouliot. "The Hows and Whys of Log Amps." *Electrical Design,* February 1, 1974, pp. 52-59.

Derivation of
Closed-Loop Responses

1. Inverting Amplifier (Fig. A-1)

At node A, the input current I_i and the feedback current I_f are related by Kirchhoff's current law:

$$I_i + I_f = 0 \qquad \text{(Eq. A-1)}$$

since no current flows into the *ideal* op-amp. Consequently,

$$I_i = - I_f \qquad \text{(Eq. A-2)}$$

By using Ohm's law, the following two equations can be written:

$$V_i - V' = I_i R_i \qquad \text{(Eq. A-3)}$$

$$V_o - V' = - I_i R_f \qquad \text{(Eq. A-4)}$$

since $I_i = - I_f$. At node A, the voltage V' is the differential input to the op-amp. Assuming this voltage to be zero*, then Equations A-3 and A-4 can be expressed as:

$$V_i = I_i R_i \qquad \text{(Eq. A-5)}$$

$$V_o = - I_i R_f \qquad \text{(Eq. A-6)}$$

The voltage gain V_o/V_i is then found by dividing Equation A-6 by Equation A-5, or,

---

* Essentially, V' is typically very small, about 1 mV. For our purpose in deriving the closed-loop response, this term is negligible. However, it should be kept in mind that V' is inversely proportional to open-loop gain.

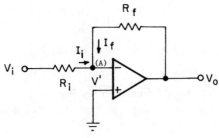

Fig. A-1.

$$\frac{V_o}{V_i} = -\frac{R_f}{R_i} \qquad \text{(Eq. A-7)}$$

which is exactly the form given by Equation 2-6. The negative sign indicates that the output is 180° out-of-phase with the input.

2. Noninverting Amplifier (Fig. A-2)

At node A, the voltage across R_i is:

$$V_A = \frac{R_i}{R_i + R_f}(V_o) \qquad \text{(Eq. A-8)}$$

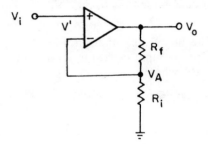

Fig. A-2.

Then, by Kirchhoff's voltage law,

$$V_i - V' - V_A = 0 \qquad \text{(Eq. A-9)}$$

Since V' is taken to be zero, Equation A-9 reduces to:

$$V_i = V_A$$

which is also equal to Equation A-8, so that:

$$V_i = \frac{R_i}{R_i + R_f}(V_o) \qquad \text{(Eq. A-10)}$$

The voltage gain is found by rearranging Equation A-10, so that:

$$\frac{V_o}{V_i} = 1 + \frac{R_f}{R_i} \qquad \text{(Eq. A-11)}$$

which is exactly the form given by Equation 2-2.

3. Differentiator (Fig. A-3)

The circuit for the differentiator in Fig. A-3 is the same as the inverting amplifier, except that the input element is now a capacitor.

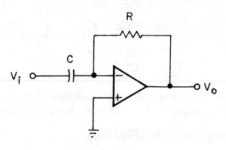

Fig. A-3.

By rewriting Equation A-7 in terms of input and feedback *impedances,* we then have:

$$\frac{V_o}{V_i} = -\frac{Z_f}{Z_i} \qquad \text{(Eq. A-12)}$$

The impedance of the feedback resistor is simply the resistance R. For the input capacitor, its impedance as a function of frequency is $1/sC$, where:

$$s = j\omega$$
$$= j2\pi f \qquad (j = \sqrt{-1})$$

Consequently the terms of Equation A-12 can now be expressed as:

$$\frac{V_o}{V_i} = -RCs$$

or the more convenient form,

$$V_o = -RCs(V_i) \qquad \text{(Eq. A-13)}$$

Since the multiplication of the input variable V_i by s implies differentiation of V_i (by Laplace transforms), Equation A-13 is then rewritten as:

$$V_o = -RC\frac{dV_i}{dt} \qquad \text{(Eq. A-14)}$$

which is exactly the form given by Equation 3-2.

4. Integrator (Fig. A-4)

By using the same approach as has been done with the differentiator,

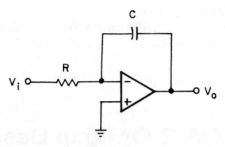

Fig. A-4.

$$\frac{V_o}{V_i} = -\frac{Z_f}{Z_i} \qquad \text{(Eq. A-15)}$$

Substituting $Z_i = R$ and $Z_f = 1/Cs$ into Equation A-15, we have after solving for V_o,

$$V_o = -\frac{1}{RC}\frac{V_i}{s} \qquad \text{(Eq. A-16)}$$

Since the division of the input voltage by the variable s implies the integration of V_i (by Laplace transforms), then Equation A-16 is rewritten as:

$$V_o = -\frac{1}{RC}\int_0^t V_i dt \qquad \text{(Eq. A-17)}$$

which is the form given by Equation 3-14.

The OA-2 Op-Amp Designer

The OA-2 Op-Amp Designer, shown in Fig. 1-7, is one of the many breadboarding aids manufactured by E&L Instruments, Inc. It is a self-contained breadboard and design station which can be used to perform almost all of the experiments given in this workbook. The OA-2 designer system includes:

- SK-10 Universal Breadboarding Socket.
- +5 volt power supply: 500 mA, current limited.
- ±15 volt power supply: 200 mA each, current limited.
- Null indicator (2 LEDs): detects less than either a 10 mV or a 100 mV difference (switch selectable).
- Function generator: generates simultaneous sine, square, and triangle waveforms with less than 3% total harmonic distortion. The frequency of the generator can be either switch selectable for a constant frequency of 500 Hz, or adjustable by the use of external capacitors.
- 2 uncommitted spdt slide switches.
- 2 uncommitted potentiometers: 10 kΩ and 100 kΩ.
- 2 BNC-type connectors.

When the OA-2 designer is used to perform the experiments in this book, a few minor changes must be recognized.

1. Use the 100 kΩ potentiometer instead of the 50 kΩ potentiometer shown in the schematics.
2. Using the 100 kΩ potentiometer on the OA-2 designer, wire the circuit shown in Fig. B-1 and connect it to the output of the function generator, thus giving a variable amplitude control.

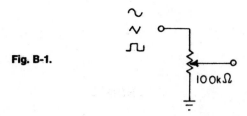

Fig. B-1.

3. The function generator has a fixed frequency of 500 Hz. However, when the *frequency* switch is in the EXT position, the frequency is user adjustable by the insertion of an external capacitor into the socket pins. The resulting frequency is determined by the approximate formula:

$$f = \frac{5 \times 10^{-5}}{C_{EXT}}$$ (Eq. B-1)

where C_{EXT} is the external capacitance in farads. The following table gives the approximate generator frequency for the indicated value of the external capacitor.

C_{EXT} (μF)	Frequency (kHz)	C_{EXT} (μF)	Frequency (Hz)
.001	50.0	.068	735
.0022	22.7	.082	610
.0033	15.2	.1	500
.0047	10.6	.22	227
.0068	7.4	.33	152
.0082	6.1	.47	106
.01	5.0	.68	74
.022	2.3	.82	61
.033	1.5	1.0	50
.047	1.1		

Since we are not able to smoothly vary the frequency of the OA-2, you will encounter some difficulty in performing the experiments in Chapter 7. For all the other experiments requiring a specific frequency, use the capacitor value that gives the frequency closest to the desired value.

Index

TO THE READER

This book is one of an expanding series of books that will cover the field of basic electronics and digital electronics from basic gates and flip-flops through microcomputers and digital telecommunications. We are attempting to develop a mailing list of individuals who would like to receive information on the series. We would be delighted to add your name to it if you would fill in the information below and mail this sheet to us. Thanks.

1. I have the following books:

_____

_____

_____

_____

_____

_____

_____

_____

_____

_____

_____

2. My occupation is: ☐ student   ☐ teacher, instructor   ☐ hobbyist

      ☐ housewife   ☐ scientist, engineer, doctor, etc.   ☐ businessman

      ☐ Other: _____

Name (print): _____

Address _____

City _____ State _____

Zip Code _____

Mail to:

        Books
        P.O. Box 715
        Blacksburg, Virginia 24060